# Springer Theses

## Recognizing Outstanding Ph.D. Research

For further volumes:
http://www.springer.com/series/8790

## Aims and Scope

The series "Springer Theses" brings together a selection of the very best Ph.D. theses from around the world and across the physical sciences. Nominated and endorsed by two recognized specialists, each published volume has been selected for its scientific excellence and the high impact of its contents for the pertinent field of research. For greater accessibility to non-specialists, the published versions include an extended introduction, as well as a foreword by the student's supervisor explaining the special relevance of the work for the field. As a whole, the series will provide a valuable resource both for newcomers to the research fields described, and for other scientists seeking detailed background information on special questions. Finally, it provides an accredited documentation of the valuable contributions made by today's younger generation of scientists.

## Theses are accepted into the series by invited nomination only and must fulfill all of the following criteria

- They must be written in good English.
- The topic should fall within the confines of Chemistry, Physics and related interdisciplinary fields such as Materials, Nanoscience, Chemical Engineering, Complex Systems and Biophysics.
- The work reported in the thesis must represent a significant scientific advance.
- If the thesis includes previously published material, permission to reproduce this must be gained from the respective copyright holder.
- They must have been examined and passed during the 12 months prior to nomination.
- Each thesis should include a foreword by the supervisor outlining the significance of its content.
- The theses should have a clearly defined structure including an introduction accessible to scientists and experts in that particular field.

Ajay Virkar

# Investigating the Nucleation, Growth, and Energy Levels of Organic Semiconductors for High Performance Plastic Electronics

Doctoral Thesis accepted by Stanford University
for Chemical Engineering, Stanford, CA, USA

 Springer

*Author*
Dr. Ajay Virkar
c3Nano Inc.
Villa Street 368
Mountain View
CA 94041
USA
e-mail: ajay@c3nano.com

*Supervisor*
Prof. Dr. Zhenan Bao
Stanford University
381 North South Mall
Rm. 213 Stanford
CA 94305
USA
e-mail: zbao@stanford.edu

ISSN 2190-5053
ISBN 978-1-4419-9703-6
DOI 10.1007/978-1-4419-9704-3
Springer New York Dordrecht Heidelberg London

e-ISSN 2190-5061
e-ISBN 978-1-4419-9704-3

Library of Congress Control Number: 2011936138

*Cover design:* eStudio Calamar, Berlin/Figueres

Printed on acid-free paper

Springer is part of Springer Science+Business Media (www.springer.com)

# Preface

Plastic or organic electronics offer several advantages over conventional inorganic technologies. Firstly, the molecular structure of organic semiconductors and conductors can be tuned for various applications using synthetic chemistry. In addition organic thin films are flexible, and can be processed and patterned inexpensively. However, improving the thin film conductivity of organic semiconductors and conductors is necessary for widespread application and adoption. *The overall goal of this thesis is to investigate and control organic small molecule growth at surfaces in order to improve the performance of organic electronic devices.*

In Part I of the thesis, improving the charge carrier mobility of organic thin film transistors (OTFTs), the building block for plastic electronics, is discussed. The nucleation, stability and thin film growth of model organic semiconductors such as pentacene and $C_{60}$ are described with focus on correlating thin film structure to charge carrier mobility. More specifically, pentacene nucleation and growth on the most common substrate for OTFTs, an octadecylsilane (OTS) monolayer modified silicon oxide surface, is investigated. The role of the density of the OTS was determined to be a critical device parameter that impacts organic semiconductor nucleation and growth, and the charge carrier mobility, as the OTS transitions from an amorphous monolayer into a crystalline one. Dense OTS monolayers were fabricated using the well known ultrathin film Langmuir Blodgett (LB) technique, as well as a new spin-coating technique developed in our lab. The crystalline OTS monolayer serves as an excellent template for promoting desirable organic semiconductor thin film growth leading to high performance transistors. Therefore a crystalline OTS dielectric surface modification layer, which greatly improves organic semiconductor performance, may be important for the future success of OTFTs and organic circuits.

In the Part II of the thesis, lessons learned from studying organic semiconductor nucleation and growth are applied to improving the conductivity of carbon nanotube (CNT) networks for transparent electrode applications. Selective growth

of organic small molecules with low molecular orbital energies was used to greatly
reduce the sheet resistance of CNT films by both decreasing junction resistances
and stable doping of the semiconducting CNTs. The result is a material which has
the highest value (in terms of transparency and sheet conductivity) of any carbon
based transparent electrode.

# Acknowledgments

I want to thank Zhenan for being an excellent adviser. Not only did I learn a tremendous amount about organic electronics, surface chemistry, material science, and transistors from her, but she taught me how to conduct independent research. She also always gave me the freedom to pursue my research interests and ideas. She is not only a world-class scientist, she is also a remarkably kind, thoughtful and caring person. She is professionally and personally someone I admire and I could not have asked for a better adviser or mentor.

I was fortunate to work with three great postdocs during my time here. Jason Locklin introduced me to the field of organic electronics and surface chemistry and helped me learn how to think about experiments. His positive attitude and enthusiasm were contagious. I have learned more about thin film growth from Stefan than anyone. He is a very gifted scientist/teacher and a great friend. He also helped me think about the basic molecular processes which were occurring during physical phenomena and taught me to think about energy and interactions when investigating the material science I was interested in. Finally, Melbs is not only a great guy, and about the hardest working guy in lab, but he is also one of my closest friends and the person who I have spent the most stressful and exciting 3 AM lab nights with. Our collaboration on carbon nanotube electronics has been fascinating and rewarding.

I also want to thank my committee for all their help along the way. Curt and Andy met with me several times and helped me refine my thinking and improve the quality of my thesis. Also, Andy helped me a lot on an initial project on liquid crystals and his Polymer Physics course is the best class I took at Stanford. Alberto, is a true expert in organic transistors and he also provided a lot of guidance, helpful ideas and suggestions. I also enjoyed collaborating with members of his group.

The Bao group (past and present) has been a tremendously fun and exciting place to work. I have made lifelong friends and have truly enjoyed my time with all of you. I especially want to thank Yutaka, Ken, Randy, Ming Lee, Soumen, Hylke, Joe, Arjan Mark, Colin, Shuhong, Maria, Justin, and Chris, (and anyone else I might have forgotten) for their help, discussion, and hilarious nonsense over the past 5 years.

All my friends at Stanford outside of lab have made things fun, and rather beer-filled. Thanks to my roommates and close friends Tom ("the Horse") and Cem ("the Turk") for the last two years and especially the past few months when I have been very busy. Also I want to thank Soumen and Randy who are really good friends and was always willing to help me out. I also had a great time Welsh, Tracy, Kedar, Krishna, Dane, Sandeep, and Bobby. Thanks for all the fun for all the help.

I have had the fortune of also having many other friends outside of Stanford. Thanks to all my boys from Illinois and Utah (Santosh, Kuki, Rytas, Mark, Mohammed) and (Jesse, Mikey, Conor, Preston, and Kosh). I want to thank Hausser one of my best friends for often challenging the way I thought. I have to thank two of my best friends, Sravan and Vivek who have always made me laugh and who have always been there for me.

I want to thank my family. Everyone in India (my eldest brother and sister Beej and Aswhini Tai, and Maushi, Kaka, and everyone else) for everything over the past 27 years. Thanks to my brother and his wife, Shashank and Smita for being supportive and caring and making sure I always have a great time when I come to Utah (and making me and uncle!). AJ and Amol are also like older brothers and I admire them both a lot. AJ especially has helped me and given me a lot of useful advice.

Most importantly I thank my parents. My dad is an incredible scientist and his integrity and passion for science are inspiring. He is still writing first author papers with 100 equations! Finally to my Mom, who will always be my favorite person, everything good I ever do is because of you. Your kindness, personality, and love have shaped me into the person I am. Thanks to you both, for all your love and support.

# Contents

# Chapter 1
# Introduction to Organic Semiconductors, Transistors and Conductors

## 1.1 Background: Plastic Electronics

The discovery that carbon based materials can efficiently conduct electricity is among the most significant findings in material science in the past several decades and spawned the field of organic or "plastic" electronics. The discovery of conducting polymers was made by physicist Professor Alan Heeger, and chemists Professor Alan MacDiarmid and Professor Hideki Shirakawa for which the three were awarded the Nobel Prize in chemistry in 2000. Plastics, along with conventional inorganic semiconductors like silicon, represent the two most significant materials breakthroughs in the past century. It would be impossible to imagine life without either. Organic electronics is a fascinating emerging branch of material science since its aim is to wed the benefits of these two classes of materials.

As Nobel Prize winner Alan Heeger stated, organic semiconductors and conductors, "offer a unique combination of properties not available from any other known materials." Compared to brittle inorganic materials, which are expensive to process and require high temperatures and vacuum, organics can be processed inexpensively, at low temperatures and are compatible with flexible plastic substrate [1]. Unlike inorganic crystalline materials where the lattice is composed of covalent or ionic bonds between neighboring atoms, in organic crystals, the lattice is held together by weaker van der Waals forces which allow the material to be flexible and ductile [2]. Finally, a wide variety of conductive organic materials can be synthesized using chemistry. The ability to tailor or tune a semiconductor or conductor for a specific application is another major advantage plastic electronics have over conventional inorganics [2, 3].

The potential to create novel, flexible, and cheap electronics, has engendered thousands of researchers worldwide to study plastic electronics. This has lead to the investigation and implantation of organic semiconductors and conductors into a variety of electronic devices some of which are now widely commercially available such as organic light emitting diode (OLED) displays, and others which

A. Virkar, *Investigating the Nucleation, Growth, and Energy Levels of Organic Semiconductors for High Performance Plastic Electronics*, Springer Theses, DOI: 10.1007/978-1-4419-9704-3_1, © Springer Science+Business Media, LLC 2012

are still being investigated such as organic solar cells. Widespread adoption of plastic electronics still requires considerable research and development. Moreover, finding the "niche" technologies where the advantages of organics can be realized is an important consideration for commercialization.

Aside from carbon nanotubes and graphene (discussed later), most traditional carbon-based semiconductors and conductors still lack the electrical conductivity to drive the logic in high end applications such as a computer processor; however their performance is sufficient for a variety of lower-speed applications, such as a computer display [1, 2]. Plastic Logic, a leading start-up company commercializing organic devices, plans to unveil its electronic-reader named "que" in 2010. It can store hundreds of books, and documents, all at a fraction of the weight of conventional inorganic technologies. The backplane of the plastic display is driven by organic circuits which utilize a semiconducting polymer as the active layer. Sony has recently developed organic light emitting diode (OLED) monitors and televisions. Currently the cost is high, but OLED displays require less power to operate, and emit uniformly (i.e., there is no angular viewing dependence like liquid crystal displays) [2]. Moreover, the energy levels of the organic semiconducting molecules in an OLED can be tuned by chemistry, so very aesthetic and "eye-pleasing" colors can be emitted. Other applications for plastic electronic materials include radio frequency identification (RFID) tags, and other flexible, low-cost electronic circuitry. If the stability and performance of organic material increase, it is possible extremely cheap RFID tags (less than $0.01) could be fabricated. Retailers hope to tag even the most common items like a bag of potato chips with RFID tags. Not only would this decrease the chance for merchandise to be stolen, but one could imagine checking out a store with all the information simultaneously transferred via these RFID tag to the consumer's debit or credit card, minimizing time and money spent using checkouts or cashiers. Tracking and supply chain monitoring of goods would also greatly benefit from cheap organic RFID tags.

In the past several years there has also been a surge of research activities on fabricating organic semiconductor based biological and chemical sensors [4]. Again the ability to tune the materials using chemistry for high analyte sensitivity/ specificity and the potential for low-cost fabrication make organics very promising. The flexibility and potential compatibility with a variety of biomaterials, also make organic electronics interesting for biomedical applications [5]. Finally, the past several years have seen a tremendous rise in the research and development of organic solar cells. The current record efficiency for organic solar cells is $\sim 8\%$ which is quite impressive. Several startups have begun to populate this area with fierce competition and impressive speed. If several key challenges are addressed, it is possible organic solar cells will be competitive with inorganic technologies in the near future (Fig. 1.1).

All the aforementioned technologies require high performance organic semiconductors and or conductors. The physical distinction between the two and a more in-depth analysis of types of electrical circuit elements which are put together for a real device are described below.

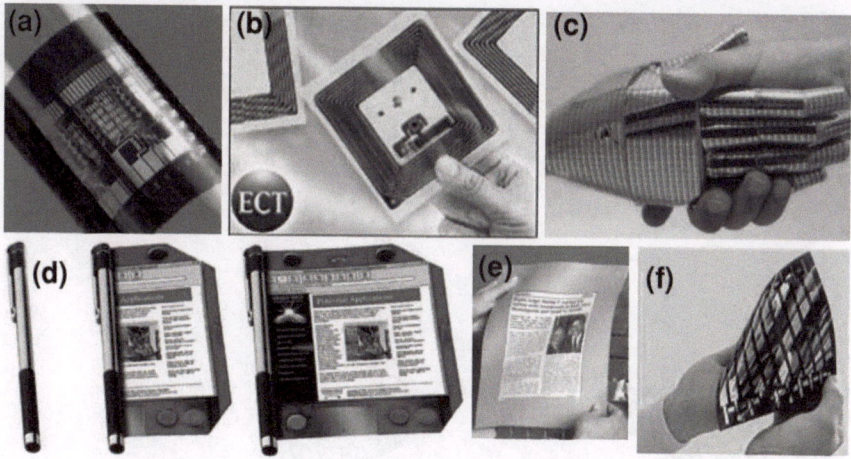

**Fig. 1.1** Organic electronic devices **a** flexible circuitry from ECT inc, **b** flexible radio frequency identification tag, **c** flexible pressure sensors made using organic transistors from Univ. of Tokyo, **d** concept of a flexible display from ECT, **e** prototype flexible electronic paper from Plastic Logic, **f**, a flexible organic solar cell from Konarka Inc.

## 1.2  Organic Semiconductors and Conductors

In order for organic materials to be semiconductors or conductors, they must possess a rich pi-electron system, i.e., molecules composed of sp$^2$ hybridized carbons like aromatics and heteroaromatics. The double bonded character allows pi-electrons in a material to be mobile and thus conduct electricity [2, 6–8].The vast majority of conventional plastic or organic materials like plastic bottles, bags, bike helmets, and plates are composed primarily of insulating polymers where the backbone of the polymer consists of single bonds between carbons. These plastics are flexible and easy to process, but are electronically insulating (the bandgap is typically in excess of ~4–5 eV). These types of materials are interesting as insulators (dielectrics) and encapsulation. Scientists working on organic electronics are primarily focused on improving the conductivity, stability, and tailorability of highly conjugated organic semiconductors (bandgaps are typically in the 1–3 eV range) and conductors [2, 9].

Several common organic semiconductors and conductors are illustrated in (Fig. 1.2). There are several features which dictate their performance and thus potential utility in devices, which will be discussed in the next two sections. Similar to inorganic materials, the terms insulator, semi-conductor, and conductor refer to the energetic differences between the lowest unoccupied molecular orbital (LUMO) or conduction band (CB) and the highest occupied molecular orbital (HOMO) or valence band (VB). For insulators the band gap is appreciably large. Upon application of an electric field there is still not enough energy for electrons to populate the LUMO or CB. Semiconductors have received the most industrial and

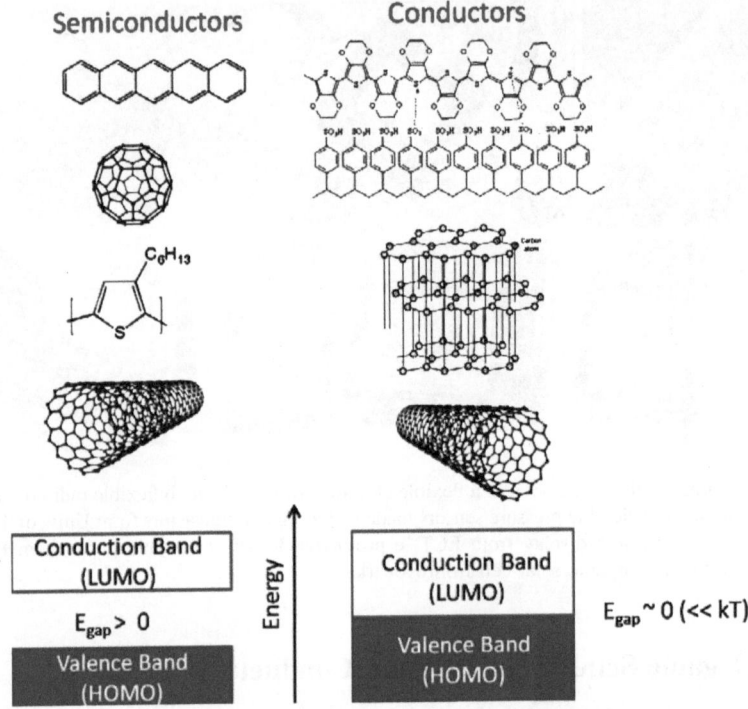

**Fig. 1.2** Common carbon-based semiconductors and conductors. Semiconductors from the *top* to *bottom*: pentacene, $C_{60}$, poly-3-hexylthiophene, semiconducting single walled carbon nanotube. Conductors from *top* to *bottom* poly-3,4-ethylenedioxythiophene polystyrenesulfonate (PEDOT:PSS), graphite, metallic single-walled carbon nanotube. (*note the $k$T refers to thermal energies where $k$ is Boltzmann's constant and $T$ is temperature)

research attention. Semiconductors are the active component in electronics since they can be operated as switches (more in the next section) [2]. Upon application of electric fields the semiconductor can conduct current like a conductor, and upon the removal of external fields, is insulting. The band gap is intermediate in energy. Finally, conductors are materials where the band gap is zero or very small compared to thermal energy [9]. Conductors are also critical for electronic applications and there are several interesting organic materials which can conduct as efficiently as metals, but are lighter and more flexible.

It is important here to note that throughout the remainder of this thesis the terms "carbon-based" and "organic" are used interchangeably. Strictly speaking, allotropes of carbon, or those which contain pure carbon like carbon nanotubes, graphene, and $C_{60}$, are technically not defined as "organic". However, in the field of plastic electronics they are commonly grouped with organic materials like those shown in Fig. 1.2, and will also be considered as organic conductors and semiconductors in this work.

## 1.3  The Organic Thin Film Transistor

The ability for the semiconductor to act as either "on" and conduct current when external energy is applied, and behave as insulator or "off" when the external energy is removed, allows for the fabrication of logic circuits. The building block of organic circuitry is the organic thin film transistor. The majority of research efforts in organic electronics have been dedicated to improving the performance of organic thin film transistors (OTFTs) [1, 2, 10–13]. The active layer is the organic semiconductor (see Fig. 1.3). Analogous to inorganic semiconductors, control of the chemistry, materials purity/quality, thin film growth, and processing dictate the performance of organic materials. Typically, three major metrics are used to assess the performance of an organic semiconductor. The first is the charge carrier mobility, $\mu$ (cm$^2$ V$^{-1}$ s$^{-1}$), which is a measure of how fast charge carriers can move (i.e. their velocity v) under an applied electric field (E) [2].

$$\mu = v/E \qquad (1.1)$$

The mobility determines the potential usage of the semiconductor. For simple circuits which can function as backplanes for displays, the mobility must be $\sim 0.1$–1 cm$^2$ V$^{-1}$ s$^{-1}$. This is the value for amorphous silicon, which is currently used to drive most displays. The modest mobility needed for display applications is due to the fact that the human eye cannot sense faster speeds [8]. The transistors in a computer processor, however, require mobilities of $\sim 500$–1,000 cm$^2$ V$^{-1}$ s$^{-1}$ to perform complex tasks quickly. This is achieved by using very high purity, crystalline silicon. Aside from carbon nanotubes and graphene (discussed later) no organic semiconductor to date shows such high mobilities. Typical mobilities for good organic semiconductors fall into the 0.1–5 cm$^2$ V$^{-1}$ s$^{-1}$ range for thin film TFTs and up to 1–30 cm$^2$ V$^{-1}$ s$^{-1}$ for single crystal TFTs [10]. Two other important metrics are the on/off ratio and the threshold voltage ($V_T$). The on/off ratio is a measure of the maximum current when the TFT is turned on (when voltages are applied) to when the device is off (no voltage applied). This value should also be as high as possible; typical values for good OTFTs are $10^5$–$10^8$. $V_T$ is essentially the voltage required to turn on the device and should be as close to zero volts as possible in order to minimize power requirements and improve switching speeds. Other significant metrics include the sub-threshold swing, hysteresis characteristics, and stability [2, 14].

The typical OTFT structure is shown in Fig. 1.3. The semiconductor is deposited onto a organosilane self assembled monolayer (SAM) atop a 300 nm SiO$_2$ dielectric layer. The insulating SiO$_2$ dielectric is typically thermally grown on top of the gate electrode-a heavily doped silicon wafer (the doping is high enough that the silicon is metallic). The importance of the SAM will be discussed in detail later. The OTFT is completed by defining gold source and drain electrodes via thermal evaporation through a shadow mask. Though many of the materials used for OTFT fabrication are inorganic (Si, SiO$_2$, gold) they are used since they are readily available in a research setting. The Si/SiO$_2$ substrate can be

**Fig. 1.3** Typical schematic of an organic thin film transistor. S and D refer to the source and drain electrodes (which are generally gold, but many other metals are also being researched)

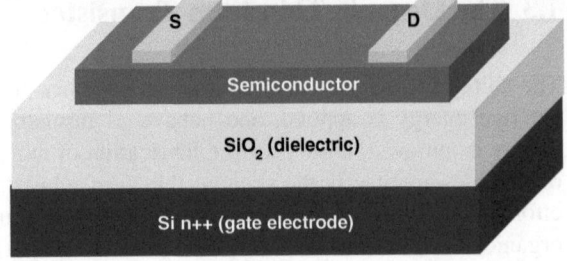

purchased in large quantities from the silicon wafer fabrication industry [2, 15]. The active organic semiconductor is the most critical for plastic electronics. Recently many groups have demonstrated all organic circuits (i.e. electrodes, semiconductor, and dielectric are all organic) [1, 16–18]. But for practical screening of organic semiconductor performance Si/SiO$_2$ remains the most common substrate.

## 1.3.1 OTFT Operation and Carrier Type

An OTFT operates in the following way: a voltage is applied to the gate electrode which induces a thin sheet of charge at the semiconductor/dielectric interface. Typically voltages of ∼100 V are dropped over the 300 nm dielectric, so that the vast majority of mobile charges are confined to a few nanometers at the semiconductor/dielectric interface [19, 20]. As discussed in detail later, control of this interface is critical for device performance. Another voltage is applied between the source and drain electrodes, which causes the mobile charges in the semiconductor to move from source to drain and thus establish a current flow [9].

Organic semiconductors can act as either electron transporters (n-channel) or hole transporters (p-channel). The carrier type is dependent on trap states, the HOMO–HOMO wave function overlap between neighboring molecules, the LUMO–LUMO wave function overlap, and the relative positions of the HOMO and LUMO with respect to the Fermi energy of the source and drain electrodes [9]. Pentacene, for example, is the most studied molecule for organic thin film transistor applications, since high hole mobilities (over $1.0 \text{ cm}^2 \text{ V}^{-1} \text{ s}^{-1}$) can be achieved regularly using vacuum deposition. Though much less commonly demonstrated, pentacene has also been used as an electron transporter. Gold is typically used as the source and drain electrodes in OTFTs, the energy barrier between the HOMO of pentacene and the Fermi energy of gold is smaller than between the Fermi energy and the LUMO of pentacene. Thus upon application of negative voltages, the energy levels of the HOMO and LUMO of pentacene are increased, until they are resonant with the Fermi energy of gold. The electrons from the HOMO spill out onto gold, leaving behind mobile holes. Upon application of a negative voltage between the source and drain, holes are free to move (Fig. 1.4).

**Fig. 1.4** Schematic showing p-channel operation of a TFT. Application of gate voltage draws charge to the dielectric interface, and then by applying a source-drain induces mobile charge to flow establishing a current. Once the voltages are removed, the semiconducting layer returns to the "off" state

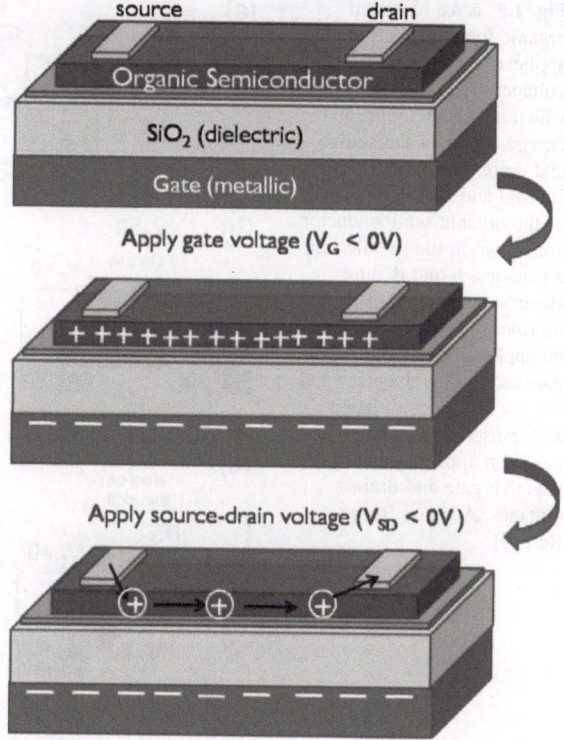

Conversely, molecules with lower HOMO and LUMO energies are often regarded as n-channel semiconductors. $C_{60}$, for example is a well known n-channel material—which is again due the fact its LUMO is closer in energy to Fermi energy of gold compared to its HOMO. Upon application of a positive gate voltage, the HOMO and LUMO energies of the $C_{60}$ lower in energy to the point where the LUMO is resonant with the Fermi Energy of gold [9]. As the energy is lowered by the applied voltages, electrons from the gold populate the $C_{60}$ LUMO, and upon application of a positive voltage difference between the source and drain electrodes, the electrons flow establishing a current (Fig. 1.5).

The charge carrier mobility can be extracted from current–voltage (*I–V*) measurements. The source-drain current is a function of the applied voltages at the gate electrode, between the source and drain, the geometry of the channel, and the capacitance of the dielectric. Thus, by fabricating a TFT the performance of the semiconductor can be gauged. Mathematically, the drain ($I_{DS}$) current is related to the mobility ($\mu$), applied gate voltage ($V_G$), channel geometry (channel width ($W$) and length ($L$)) and capacitance of the dielectric ($C$) by: [1, 2]

$$I_{DS} = \frac{WC}{2L}\mu(V_G - V_T)^2 \qquad (1.2)$$

**Fig. 1.5  a** An idealized organic transistor before application of external voltages. The large rectangles which are shaded represent energy band, for the source and drain electrodes. The HOMO and LUMO energies of the organic semiconductor are shown in the middle thin rectangles. **b** and **d** show electron accumulation and transport (n-channel) based on applications of positive gate and drain voltages. **c** and **e** show hole accumulation and transport (p-channel) based on applications of negative gate and drain voltages. Adapted from Ref. [9]

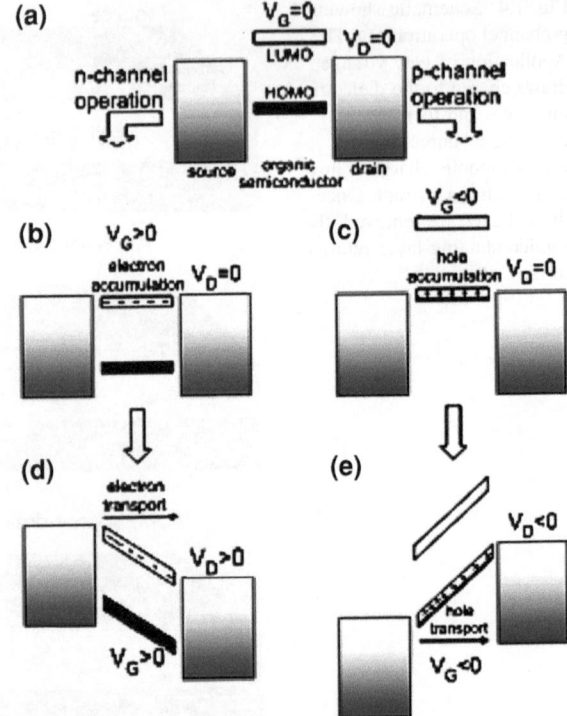

## 1.3.2 Charge Carrier Mobility

Recall that the higher the charge carrier mobility, the greater potential utility for the semiconductor. There are several factors, at various length scales, which govern the charge carrier mobility of an organic semiconductor. Considering the most commonly used microscopic theory, the rate of charge transfer between neighboring semiconducting molecules can be described by Marcus Theory [2],

$$K_{ET} = \frac{4\pi}{h} \frac{1}{\sqrt{4\pi k_B T \lambda}} t^2 \exp(-\frac{\lambda}{4k_B T}) \tag{1.3}$$

where $k_{ET}$ is the rate of electrons transfer, $h$ is plank's constant, $k_B$ is Boltzmann's constant, $t$ is the transfer integral, and $\lambda$ is the reorganization energy. The transfer integral is related to electronic overlap between adjacent molecules. The greater the overlap, the more efficiently electrons are transported. The reorganization energy is the energetic penalty associated with charging a molecule; specifically, it is the difference in energy between the charged and neutral species. The greater the reorganization energy the less efficient the charge transport [1]. There is still significant debate about how to address reorganization energies. It is also not clear how an excited molecule's reorganization energy is dissipated. There is evidence that the excited state can couple into phonon-modes of the lattice and influence many other molecules [8]. Thus the current theoretical treatment, which usually involves using density functional theory (DFT) to calculate the dimer overlap integral, and single molecule excited state energy may be far from complete. Though for many rigid small molecules, like pentacene, the treatment seems to give reasonable results [2].

However, in many cases, using quantum simulations to estimate the mobility of a thin film composed of millions of molecules leads to grossly erroneous results. The mobility measured is a macroscopic quantity, which is intimately related to the intermolecular charge transport but is also a function of several other factors. Taking a step back from the purely molecular nature of hopping between molecules as the dominant feature for charge transport, thin films are better addressed by investigating larger scale effects. Considering a thin film, there are three major categories of trap states that decrease charge carrier mobility. The first category is simply an impurity, or external molecule/material, within the thin film (or at the dielectric interface for OTFTs) which has energy levels that can trap electrons or holes. For hole trapping, the impurity may have an unoccupied state above the HOMO of the organic semiconductor; for electron trapping, the impurity may have an unoccupied state below the LUMO of the organic semiconductor. The second type of trap state involves some general distortion to the molecular packing of the molecules within a grain. This may be due to an impurity, or it may be due to some misalignments/mispacking of the organic semiconductor molecules during nucleation and growth. The misalignment of molecules within the grains hinders electronic overlap (see Eq. 1.3), which greatly decreases the charge transport. The third type of trap state, and often the most detrimental, is the grain boundary. The active channel in the TFT is composed of several grains which form a poly-crystalline thin film. Typically, despite the presence of impurities and misalignments within a single grain, the grain boundaries still dominate electrical performance. In order for a charge to traverse a boundary there is a considerable energetic penalty [1].

The simplest model, derived for amorphous or highly polycrystalline silicon, which works well for high mobility organic semiconductor thin films, focuses on the grain boundaries. The effective mobility ($\mu_{eff}$), the mobility measured, is assumed to be related to the intragrain mobility ($\mu_g$) and the grain boundary mobility ($\mu_{gb}$), and the effective length of the grains (L), the grain length ($L_g$) and the grain boundary length ($L_{gb}$) following Eq. 1.4: [1] (Fig. 1.6).

$$\frac{L}{\mu_{eff}} = \frac{L_g}{\mu_g} + \frac{L_{gb}}{\mu_{gb}} \tag{1.4}$$

However, in many cases since grain mobility is much larger than the grain boundary mobility $\mu_g \gg \mu_{gb}$, and the length of the grain is also much larger than the length of the grain boundary, Eq. 1.4 can typically be loosely approximated (1.5):

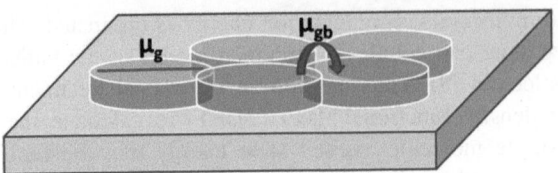

**Fig. 1.6** The two components of the effective (measured mobility) as approximated by the intragrain mobility ($\mu_g$) which is a measure of how fast charge carriers move within a grain and the grain boundary mobility domains ($\mu_{gb}$) which is a measure of fast charge carrier can move between grains

$$\mu_{\text{eff}} \approx \mu_{gb} \qquad (1.5)$$

Thus, the severity of the grain boundaries has a profound influence on the charge carrier mobility. The number and severity of the grain boundaries is intimately related to the growth and crystalline order of the semiconducting thin film and will be discussed in the next several sections [2].

## 1.4 The Dielectric/Semiconductor Interface in Organic Thin Film Transistors

As aforementioned, during TFT operation, the gate electric field draws the vast majority of mobile charges to within the first few molecular layers of organic semiconductor at the semiconductor/dielectric interface. Thus, the growth and crystalline order of these interfacial semiconducting layers can dictate transistor performance (a theme which will be repeated several times throughout this thesis). By controlling the semiconductor thin film growth and minimizing detrimental grain boundaries, high charge carrier mobilities can be achieved [19, 21–23]. In order to improve performance, or to pattern the semiconductor, the SiO$_2$ dielectric surface is nearly always modified with a self assembled monolayer (SAM), or replaced with an organic polymeric dielectric [15, 24–31]. The effect of surface modification can be dramatic since the growth, crystalline order, and even electronic properties of the semiconductor are very sensitive to the dielectric/semiconductor interface [16, 32–37] (Fig. 1.7).

The aim of the following several sections will be to highlight some general considerations about heterogeneous nucleation and thin film growth modes of organic semiconductors at the dielectric interface and methods to control preferential growth modes to achieve high performance transistors. The effects of deposition conditions on nucleation and thin film growth will be also described. Finally, as a general background to majority of the work contained in this thesis, the use of SAMs to control semiconductor growth behavior will be introduced.

**Fig. 1.7** *I–V* data from
Ref. [19] showing that due to
the applied gate field the
drain current saturates at the
~3rd monolayer in a
pentacene TFT

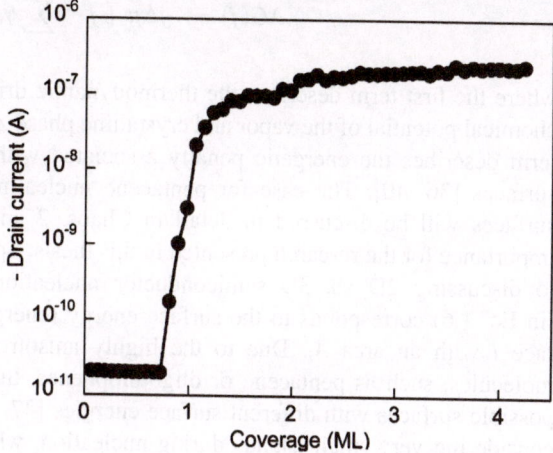

## 1.4.1 Organic Semiconductor Nucleation and Growth on Dielectric Surfaces

Typical organic transistors are fabricated either by vapor or solution deposition. In the first case, the semiconductor is vaporized and then condenses into a film onto a substrate (again typically a heavily doped silicon wafer with 200-300 nm of insulating $SiO_2$) [2, 6]. For solution deposition, the organic semiconductor is dissolved in an organic solvent. The solution is then cast onto the substrate by either drop-casting, spin-coating, dip-coating or printing [1, 2, 8, 11]. As the solvent vaporizes, the solution becomes supersaturated and forms organic semi-conductor crystals. The nucleation and growth dynamics for solution deposited organic semiconductors are more complex than vapor deposited semiconductors, since one must consider: solvent/vapor interactions, solvent/substrate interactions, solute/solvent interactions, and solute/substrate interactions [38]. Since the same thermodynamic arguments apply for both vapor and solution deposition, the next section will focus on vapor phase nucleation and growth for simplicity.

The thermodynamic driving force for nucleation is the difference between the chemical potential of the organic molecules in the vapor phase ($\mu_v$) and the crystalline phase ($\mu_c$). Thermodynamically, for nucleation to occur (without a heterogeneous substrate to catalyze crystal formation) the chemical potential of the vapor phase must exceed that of the crystalline phase—i.e., the vapor must be supersaturated. Nucleation of a stable crystal is associated with the formation of solid surfaces and their corresponding surface free energies [36, 39]. Nucleation is a competition between the thermodynamic driving force (volume effects and enthalpic lowering of free energy by beneficial intermolecular interactions), and the energetic penalty associated with surface effects (creation of new surfaces) The general equation for the free energy needed to form a finite-sized crystal composed of *j* molecules can be described by:

$$\Delta G(j) = -j\Delta\mu + j^{2/3} \sum_i \gamma_i A_i \qquad (1.6)$$

where the first term describes the thermodynamic driving force (the difference in chemical potential of the vapor and crystalline phase $\Delta\mu = \mu_c - \mu_v$) and the second term describes the energetic penalty associated with creating or adding to new surfaces [36, 40]. The case for pentacene nucleating and growing on different surfaces will be discussed in detail in Chaps. 2 and 3 since it is of particular importance for the research presented in this thesis. Specific attention will be given to discussing 2D vs. 3D semiconductor nucleation and growth. The term $\gamma_i$ (in Eq. 1.6) corresponds to the surface energy (energy/area) associated with surface $i$ with an area $A_i$. Due to the highly anisotropic shape of many organic molecules, such as pentacene or oligothiophenes, there are often many different possible surfaces with different surface energies [37, 40, 41]. Theoretically, when considering very small islands during nucleation, where islands can either disintegrate or grow, Eq. 1.6 can become intractable. This is due to complexities which arise in both terms in Eq. 1.6. At very small cluster sizes where nucleation is occurring, $G(n)$ is not a smooth, differentiable function. Rigorously, the Gibbs free energy $(G)$ represents a macroscopic ensemble quantity [36]. Nevertheless, Eq. 1.6 gives the correct macroscopic relationship between total free energy, crystal size, and surface energies and is a reasonable approximation for analysis of behavior during nucleation. For a fixed deposition rate and substrate temperature, the chemical potential term in Eq. 1.6 can be determined from experimentally measurable quantities such as the semiconductor's vapor pressure, and enthalpy of sublimation [37, 42]. Differences in free energy and growth modes can then be attributed to the influences of the surface and the relevant interfacial energies.

In general, the barrier to nucleation $\Delta G^*$ can be solved by setting the derivative of Eq. 1.6 with respect to the number of molecules $j$, to zero:

$$\left(\frac{\partial \Delta G(j)}{\partial j}\right)_{T,P} = 0 \qquad (1.7)$$

At $j = i$, $\Delta G(j) = \Delta G(i) = \Delta G^*$, where $i$ is the critical cluster size describing the size at which the addition of one more molecule stabilizes the cluster [36, 37, 39, 43, 44]. Thermodynamically $\Delta G^*$ represents the barrier to nucleation where the surface energy effects are greatest. Addition of more molecules to the cluster increases the intermolecular enthalpic interactions and lowers the total energy, and thus the intermolecular effects dominate the surface effects creating a stable island [36].

For a typical pentacene deposition (where the pressure in the vapor phase is $\sim 10^{-6}$–$10^{-7}$ torr, and the substrate temperature onto which the pentacene is deposited is between 50 and 90 °C and the rate of deposition is $\sim 0.1$–$0.4$ Å s$^{-1}$), the chemical potential difference driving nucleation is less than 0.08 eV [36, 37, 42] and thus thermodynamic models are valid for treating nucleation and growth However, several kinetic and scaling models adopted from classic nucleation theories developed for inorganic materials have also been applied to model pentacene growth [6, 40, 45–51]. A generalized theory about kinetic Equations for nucleation was suggested by Zinsmeister and covered in the review by Ruiz et al. on pentacene growth [1, 6, 52, 53].

$$\frac{dN_1}{dt} = F - \frac{dN_1}{d\tau} - 2U_1 - \sum_{i=2}^{\infty} U_i \qquad (1.8)$$

$$\frac{dN_i}{dt} = U_{i-1} - 2U_i (i \geq 2) \qquad (1.9)$$

$F$ is the flux of depositing molecules, $N_1$ is the concentration of organic semiconductor molecules (in units of molecules per unit area), and $N_i$ is the concentration of clusters with a critical number $i$ of molecules. $U_i$ is the rate at which diffusing molecules are captured by a cluster comprised of $i$ molecules, and $\tau$ the average time a molecule spends on the surface. This model is rather simple, because it does not account for the energetics of nucleation. Nevertheless, several groups have shown that kinetic arguments can be used to model the growth of pentacene islands, and is a valid approach when the barrier to nucleation is small [34, 50].

## 1.4.2 Rate of Nucleation

The rate of nucleation of stable crystals of organic semiconductors is a function of the rate of deposition, the substrate temperature, surface properties of the substrate, intermolecular-interactions, and molecule-surface interactions. The molecule-surface interaction terms are dependent on the physical processes occurring at the semiconductor/dielectric interface. The following energetic terms are important for heterogeneous nucleation and thin film growth: the energetic barrier to diffusion ($E_{diff}$), the energetic barrier to desorption ($E_{des}$), and the thermodynamic barrier required to form a stable island $\Delta G^*$ (which has already been discussed) [36, 39, 40]. Considering the three energetic terms, the nucleation density ($N_D$) of stable islands is given by Eq. 1.10:

$$N_D = R^\alpha \exp\left(\frac{E_i}{kT_s}\right) \qquad (1.10)$$

where $R$ is the rate of deposition, $\alpha$ is a constant related to the critical cluster size, $k$ is Boltzmann's constant, $T_s$ is the substrate temperature, and $E_i$ is the crystal disintegration energy (approximately equal to negative of the crystal formation energy for systems with a low driving force for crystallization) [36]. Assuming that the relevant energetic barriers to nucleation scale equivalently with the deposition rate (i.e. each has the same exponent), and the chemical potential driving force is small (i.e. on the order of thermal energy $kT$), then $E_i = (-E_{des} + E_{diff} + \Delta G^*)$ and Eq. 1.10 can be re-written as Eq. 1.11: 36, 39

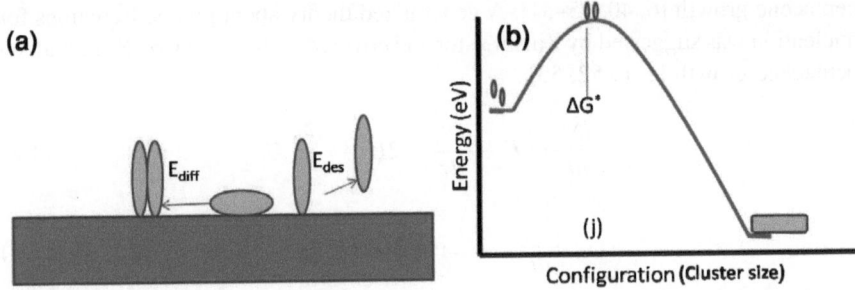

**Fig. 1.8 a** The schematic processes and energetics related to nucleation. $E_{\text{diff}}$ is the barrier required for molecular diffusion, and $E_{\text{des}}$ is the barrier for molecular desorption. **b** a schematic of the change in free energy associated with nucleation from the gas phase. The difference in free energy of the vapor phase and crystalline phase drives nucleation, however there is an activation barrier associated with nucleation; only beyond a critical cluster size do volume effects dominate surface effects forming a stable island

$$N_D = R^a \exp\left(\frac{-E_{\text{des}} + E_{\text{diff}} + G^*}{kT_s}\right) \tag{1.11}$$

It is thus evident that the three energetic barriers directly determine the nucleation density. The larger the barrier is for molecular diffusion ($E_{\text{diff}}$), the lower the surface motility and, thus, the lower the chance for molecules to encounter one-another to form a stable cluster. In general, the barrier to diffusion is the lateral corrugation in the molecule–substrate interaction potential which is determined by surface chemistry and roughness. The more tightly bound the molecule is to the surface, the greater the barrier for desorption ($E_{\text{des}}$), and thus the more time it is has to form n-mers with other molecules and finally form a stable cluster. It is clear that each of the molecular processes (i.e., diffusion, desorption, and nucleation) is also a function of the molecule–substrate interaction strength [36, 39]. It is somewhat counterintuitive, but many of the best performing organic semiconducting thin films (pentacene and $C_{60}$) showed high nucleation densities and small islands. This observation has also been made by several research groups [14, 15, 42, 54] (Fig. 1.8).

### 1.4.3 Organic Semiconductor Growth Mode

Aside from the rate of nucleation, the growth mode of the organic semiconductor is of vital significance for high performance OTFTs. Typically higher mobility semiconducting films exhibit 2D-like layer-by-layer (Frank-van der Merwe type) growth as compared to the less favorable 3D (Volmer-Weber type) growth [34, 36, 37, 42]. Three-dimensional growth at the dielectric-semiconductor

Layer-by-Layer          Stranski-krastanov          Island Growth

**Fig. 1.9** The three most common growth modes. 2D layer-by-layer (Frank-van der Merwe type) growth, A 2D monolayer followed by 3D growth (Stranski–Krastanow) and 3D island growth (Volmer-Weber type)

interface can give rise to voids in the film, and many severe grain boundaries [6, 37, 42, 55]. The growth mode is determined by a competition between inter-layer interaction energies and molecule–substrate interaction energies [36]. The stronger the molecule–substrate interactions, the greater the tendency for 2D growth. Markov showed that for 2D growth (for a simple system where the crystal is composed of cubes) to be possible, Eq. 1.12 has to be satisfied [36]:

$$\Delta\mu \geq \psi_{\text{interlayer}}\psi_{\text{mol}-\text{sub}} \tag{1.12}$$

where $\psi_{\text{interlayer}}$ is the interlayer interaction energy between an organic semicon-ductor molecule and a layer of existing organic semiconducting molecules on the surface, and $\psi_{\text{mol}-\text{sub}}$ is the interaction energy between the organic semiconductor molecule and the substrate. It is important to note that the interaction between each type of semiconductor and the surface is unique, and that simplifications based purely on hydrophobicity, surface energy, or surface chemistry arguments are incomplete [56, 57] (Fig. 1.9).

### 1.4.4 Effects of Deposition Parameters

From the preceding sections on nucleation and growth, it is evident that the physical phenomena are quite complex and there are several factors which influ-ence organic semiconductor thin film formation. Pentacene has been established as the archetypal semiconductor to study the effects of various deposition parameters, [6, 26, 28, 33, 34, 47–50, 57–61] though some groups have also investigated the growth of thiophene and naphthalene derivatives in detail [62–70]. The most common parameters are deposition rate, substrate temperature, substrate surface chemistry and surface roughness. The effect of deposition rate and the substrate temperature can be clearly seen from the analysis of Eqs. 1.10 and 1.11, which indicate that an increase in the deposition rate increases the rate of nucleation since more molecules can interact to form a stable cluster in a defined area per unit time. The effect of deposition rate has been studied in detail experimentally and indeed increasing the rate increases the nucleation density and leads to smaller crystallites [6, 40, 51, 61, 71]. The influence of the incident kinetic energy of vapor molecules on film growth and nucleation has been studied both experimentally and theoretically [72, 73].

Drastically increasing the deposition rate greatly increases the tendency for supersaturation (i.e., increases the differences in chemical potential between the vapor and crystalline phases) and the equations modeling nucleation events using purely thermodynamic arguments tend to fail. Subsequently, understanding thin film growth using analytical expressions becomes complicated. If the rate is extremely high, growth becomes completely kinetically dominated and amorphous glass-like films are formed [21]. Decreasing the deposition rate should also decrease the nucleation density and increase the overall grain size. This observation has been made by several research groups [15, 22, 42].

Assuming that the Gibbs free energy of nucleation is only a weak function of substrate temperature in the range typically used for deposition, it is also clear that increasing the substrate temperature decreases the overall barrier to heterogeneous nucleation since each term in Eq. 1.11 follows an Arrhenius-type law [36, 43, 74]. As the substrate temperature increases, so does the surface diffusion which allows for adsorbed molecules to find the lowest energy sites and form more circular (less fractal) grains [22, 35, 42, 54, 75]. The exact functional dependence between the free energy of nucleation and substrate temperature is difficult to determine. Increasing substrate temperature in general results in a decrease of the density of the semiconductor molecules on the substrate and thus they should favor the gas phase. In the extreme case when the substrate temperature is above the sublimation temperature, nucleation is thermodynamically impossible. In practice, substrate temperatures of 50–90 °C are commonly used for pentacene TFT fabrication where the vacuum pressure is $10^{-5}$–$10^{-7}$ torr [6, 37, 42]. This temperature range allows for formation of highly crystalline films which typically show good crystallinity and high mobility. Increasing the substrate temperature beyond this range appreciably decreases the sticking coefficient of the semiconductor molecule, and nucleation does not occur. Within the range where the sticking coefficient is high enough for nucleation, increasing the substrate temperature decreases the change in chemical potential between the vapor phase and crystalline phase. This increases the tendency for 2D growth since the overall supersaturation decreases (i.e., the difference in $\mu_v - \mu_c$, see Eq. 1.12). Very low substrate temperature studies (<5 °C) have also been conducted [8, 50, 71]. In such studies the diffusivity or surface mobility of pentacene molecules was very low, and there was not sufficient time or thermal energy for the molecules to find the thermodynamically favorable crystal packing, and an amorphous film was usually formed [7, 61, 82].

To summarize, for desirable grain morphology and 2D growth, the substrate temperature should be kept as high as possible, provided that desorption does not dominate and that nucleation is still possible. Finally, the majority of the substrate temperature and deposition rate studies have been conducted on bare $SiO_2$ dielectric surfaces. The influence of deposition rate and temperature on thin film growth and nucleation density can be significantly different if a SAM is used to modify the $SiO_2$ surface [36, 37, 42] or if a polymeric dielectric is used, which are discussed in the following section.

### 1.4.5   The Effects of the Dielectric Material and Dielectric Surface Modification Layer

Several materials requirements must be considered when choosing the gate dielectric. The gate dielectric should provide a trap-free interface, with a high capacitance so that an appreciable amount of mobile charges can be induced in the organic semiconductor when the transistor is turned on [1, 13, 76, 77]. The gate dielectric should also prevent gate leakage. Finally, the role of the gate dielectric on influencing the organic semiconductor crystallinity and morphology must be considered, which is the focus of the majority of this thesis and will be discussed in detail in later chapters. Comprehensive reviews on gate dielectric materials in OTFTs can be found elsewhere [2, 16, 24].

As aforementioned, due to the ready availability of silicon wafers and its surface smoothness, the most commonly used substrate for evaluating organic semiconductors is heavily doped silicon wafer as the gate electrode with a silicon oxide dielectric layer. Treating $SiO_2$ with an alkylsilane SAM can greatly improve the performance of both solution processed and vacuum deposited organic semiconductors [2]. Modifying the $SiO_2$ dielectric with an organo or alkyl-silane SAM effectively changes the surface on which the organic thin films grow and nucleate from inorganic to organic. The SAM modification also decreases hydroxyl groups from the $SiO_2$ surface, and improves the growth behavior for a variety of organic semiconductors [13, 15, 77]. The use of octadecylsilane (OTS) modified $SiO_2$ dielectrics for pentacene OTFTs can result in mobilities greater than 2.0 cm$^2$ V$^{-1}$ s$^{-1}$, whereas typical mobilities on unmodified $SiO_2$ are 0.01–0.1 cm$^2$ V$^{-1}$ s$^{-1}$ [21, 78, 79]

### 1.4.6   Controlled Growth by Surface Roughness, Surface Patterning and Surface Chemistry

The effect of dielectric surface roughness has been studied in detail [1, 2, 6, 10, 15, 28, 33, 55, 56, 76, 80–84]. Rough surfaces increase the coordination between a depositing molecule and the surface. The rough areas decrease the barrier for heterogeneous nucleation and increase the barrier to desorption [28, 36, 43, 76, 80, 84]. On rough surfaces, it is more likely that the grains are less oriented in the thin film. Thus, on very rough surfaces, the grain size is often very small, and the grains can be severely mis-oriented leading to poor performance devices [15]. However, the preferential nucleation of organic semiconductors on rough surfaces can be used to selectively grow or pattern organic semiconductors and may be a promising way to fabricate large arrays of single crystal based transistors. Recently, Briseno et al. showed that single crystal arrays of organic semiconductors can be fabricated by intentionally patterning rough pillars of OTS [10, 28]. By controlling the size of OTS pillars in the stamped regions, the number of single crystals in a

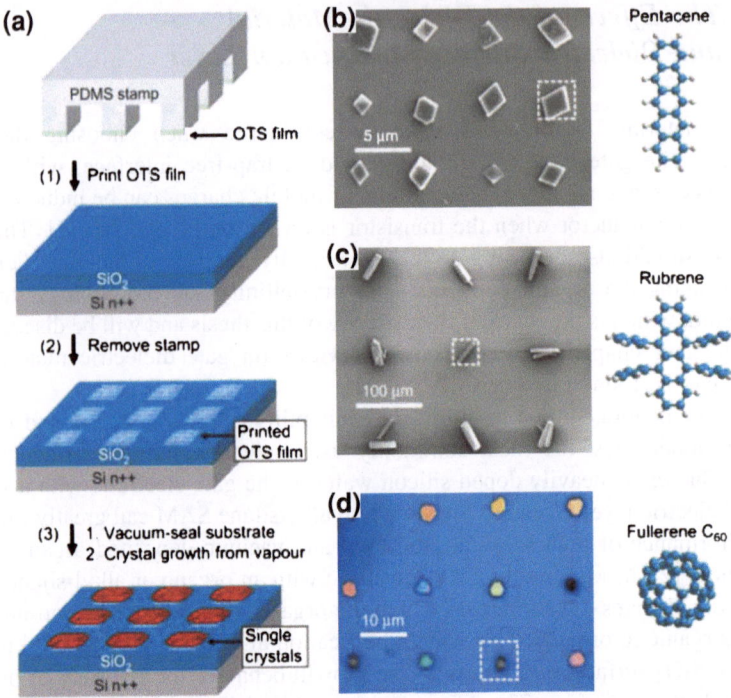

**Fig. 1.10** Single crystal arrays of organic semiconductors fabricated by patterning of rough layers of octadecylsilane layers. The pattern was deposited using soft-lithography. The high-coordination and lower barrier to nucleation promotes growth at the rough areas. Reproduced with permission from Ref. [110]

certain region could be patterned. For more detailed information on single crystal patterning see the recent review by Liu et al. [85] (Fig. 1.10).

### 1.4.7 Summary: The Dielectric/Semiconductor Interface

The most important ideas from the preceding sections can be summarized as:

1) 2D semiconductor growth is desirable for high mobility, 2) for vapor deposited semiconductors the growth mode is intimately related to the semiconductor/dielectric interaction energy, 3) controlling the nucleation and growth can achieved by considering the relevant energetics, 4) surfaces that are rough and or have high interaction energy energies readily catalyze heterogeneous nucleation and thus the growth of semiconductors can be preferentially controlled.

Chapters 2–6 will discuss the effects of the self-assembled monolayer dielectric modification on semiconductor nucleation, growth, and performance in organic thin film transistors. Chapter7 focuses on how the growth and molecular energy levels of organic semiconductors deposited on carbon nanotubes can be used to fabricate higher performance carbon-based electrodes. The following section introduces organic conductors, and specifically transparent conductors which are currently a topic of tremendous interest.

## 1.5  High Conductivity Carbon: Graphene and Carbon Nanotubes

Currently graphene and carbon nanotubes (CNTs) are among the most studied and interesting electronic materials. Graphene (see Fig. 1.11) is a 2D sheet of $sp^2$ hybridized carbons—a sheet of interconnected benzene rings. Graphene can actually be separated from graphite using scotch tape, but more sophisticated chemical vapor deposition techniques are being employed to form larger area films. Graphene is the "ultimate" 2D material, and the only carbon-based 2D crystal (1-atom thick) which has been discovered. The incredible in-plane ordering and electron delocalization makes graphene an amazing material. Mobilites as high as 200,000 $cm^2$ $V^{-1}$ $s^{-1}$ have been demonstrated in graphene (more 200 times the mobility of single crystal silicon)! [86]. Many researchers are trying to understand the exact relationship between the shape, processing and patterning of graphene in an effort to one day replace silicon as the active component in microprocessor chips [87]. The 2D highly conjugated molecular structure also gives rise to interesting physics regarding the band structure of graphene, and this is currently among the hottest topics in condensed matter physics research. In terms of real world applications, many scientists believe there a is bright future, but a better understanding graphene fundamentals and processing are still necessary [86] (Fig. 1.12).

Carbon nanotubes (CNTs) are slightly more recent in terms of research and "discovery." They can be formed from carbon soot and actually have been in use for thousands of years. CNTs were used to strengthen the Damascus sword and as ink and mascara by the Egyptians. CNT are an extremely exotic material. Structurally a CNT is a wrapped up piece of graphene (see Figs. 1.11 and 1.13). CNTs also have many incredible properties. The precise structure in which the CNT is "rolled" also called its' chirality, can give rise to either semiconducting or conducting behavior. A good approximation for a statistical average of CNTs produced by one of the many techniques is about 2/3 of the tubes are semiconducting and 1/3 are metallic. During synthesis it is also possible for several concentric CNTs to form and these are called multi-walled CNTs. Research on multi walled CNTs is primarily focused on composite or high strength materials. Single walled CNTs are more useful for electronic applications. Therefore, the discussion for the remainder of this thesis focuses on single walled CNTs.

Single walled CNTs can have aspect ratios greater than 1,000, and long-range conjugation along the tube gives rise to excellent conductivities (very large mean

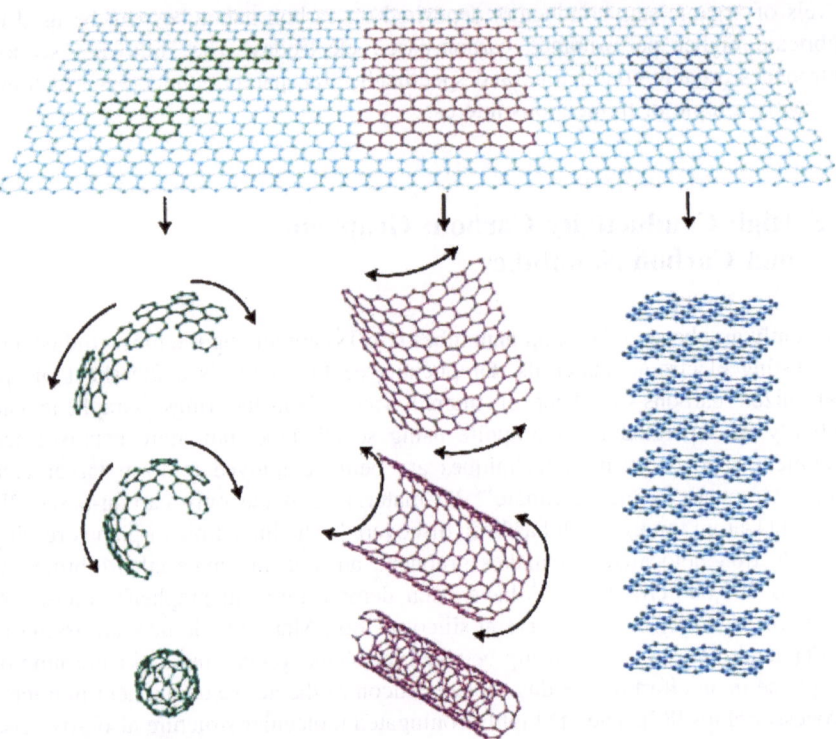

**Fig. 1.11** The various allotropes of conjugated carbon. The *top* is a sheet of graphene. The *bottom* left shows schematically that a section of 60 carbon atoms from graphene can be used to form a $C_{60}$ (fullerene). The *bottom middle* shows how a sheet of graphene can be rolled into a carbon nanotube. The *bottom right* shows the structure of graphite which is composed of stacks of graphene [86]

free paths). A single semiconducting CNT can have charge carrier mobilites in the 10,000–100,000 $cm^2$ $V^{-1}$ $s^{-1}$ range (again more than 100 times better than single crystal silicon) [88]. The conductivity of a metallic CNT can be greater than copper! This has also spurned research focused on using CNTs as a potential silicon replacement in computer processing and chips. Most of this research focuses on single CNTs placed between source and drain electrodes to fabricate high performance transistors. Currently these approaches are very expensive, and difficult to scale or pattern. The focus of the research in this thesis is on modifying the electrical properties of CNT networks. These networks are formed by depositing many CNTs to form a percolating network. CNT networks are interesting for large area applications and are much cheaper to process than single CNT architectures, though many of the exceptional properties of CNTs are also reduced. More on CNT networks will be discussed in detail in Chap. 7.

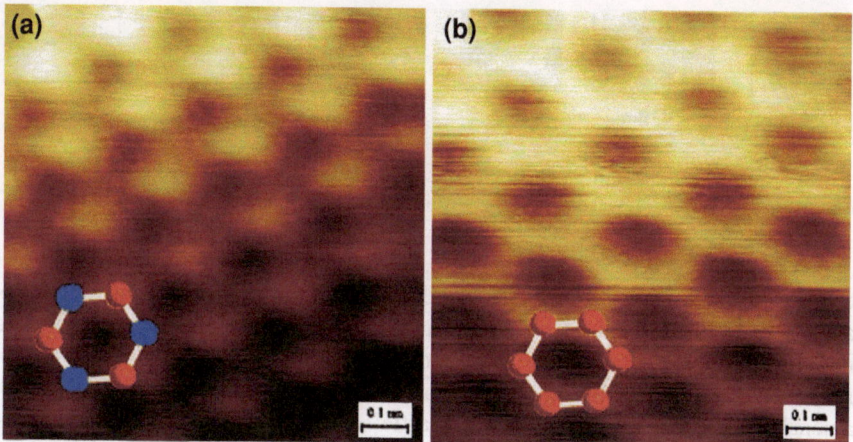

**Fig. 1.12** A high resolution scanning tunneling microscopy image of sheet of graphene. The highly regular structure composed of benzene rings can be seen [86]

**Fig. 1.13** The precise structure of the CNT determines its electronic properties. The three major classes of CNTs are show above again as being schematically "cut" from a sheet of graphene. The precise rolling (along with diameter, defects and strain) determine the band-gap of the CNT. As a rough estimate, 2/3 of CNTs are semiconducting (i.e. have a bandgap) and 1/3 are metallic (i.e., there is no bandgap)

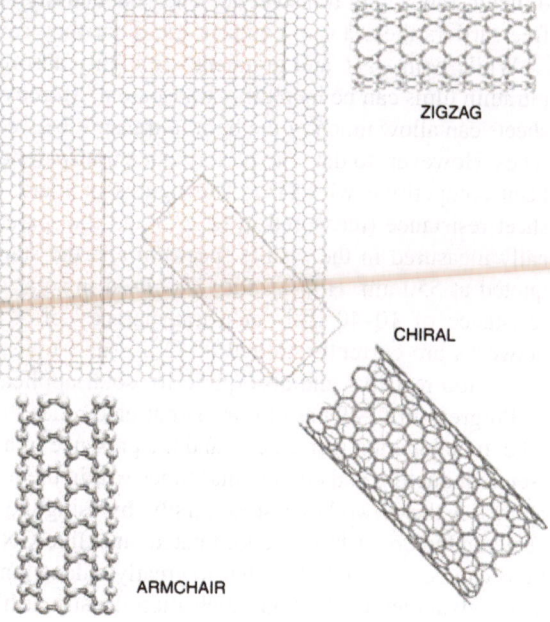

# 1.6 The Transparent Electrode

Both CNT networks and graphene films are being investigated as materials for transparent electrodes. Transparent electrodes are essential components in a variety of electronic devices—all devices where light transmission through the

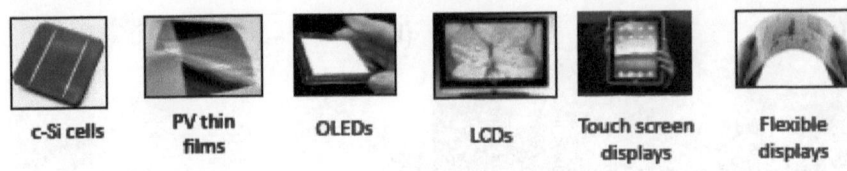

| c-Si cells | PV thin films | OLEDs | LCDs | Touch screen displays | Flexible displays |

**Fig. 1.14** The various devices/applications which require transparent electrodes

electrode is necessary. Several different devices that require transparent electrodes are shown in Fig. 1.14. Currently the materials used for transparent electrodes are mainly inorganic metal oxides. The most common is indium tin oxide (ITO). However, ITO is expensive, and due to its high processing temperatures is not amenable for processing on low-cost plastic substrates. For thin film solar technologies, for example, the ITO transparent conductor can be up to 25–40% of the entire device cost. Finally, ITO and other metal oxides, are inherently brittle. This makes them incompatible with next generation flexible electronics. ITO has remained the material of choice, because finding suitable replacements has been difficult. Physically transparency and conductivity are inversely related. The more free carriers there are to conduct, the more they can also absorb light.

While both CNT and graphene are highly absorbing (both are black in color), ultrathin films can be incredibly transparent. Moreover for a CNT network, thin 2D sheets can allow much of the light to transmit between the interstitial area between tubes. However, to date the performance transparent of CNT networks has still not been competitive with ITO. The figures of merit for transparent electrodes are sheet resistance (measured in $\Omega\square^{-1}$) and transparency. The transparency is typically measured in the visible spectrum and the standard value of transparency is quoted at 550 nm. High quality ITO used in solar cells and displays, has a sheet resistance of 10–40 $\Omega\square^{-1}$ at a transparency of $\sim$80–90% [89]. The best CNT networks are currently around 80 $\Omega\square^{-1}$ at a transparency of 80%—still considerably too resistive and absorptive for solar applications [90].

Progress in CNT based transparent electrodes (TEs) has been hindered by the tube–tube junction resistances, and the presence of a mixture of different chiralities (semiconducting and conducting) tubes within the network. Chapter 7 of this thesis addresses these two key issues. Firstly, by using the preferential growth of organic semiconductors at highly coordinated sites like CNT–CNT junctions, these junctions can be "welded" together to greatly reduce junction-contact resistances. This takes advantage of the high nucleation density at high energy sites. Furthermore, due to the stronger interaction energy the small molecule organic semiconductors have with the CNTs compared to underlying substrate (glass, plastic, of $SiO_2/Si$), the organic semiconductor growth is primarily on the CNT network, so that the transparency is not reduced (i.e., the interstitial area remains free of additional material). Finally by selecting organic molecules with Fermi energies lower than the Fermi energies of the semiconducting CNTs in the networks, the semiconducting tubes are doped which also greatly increases conductivity.

## 1.7 Conclusion and Goals of the Thesis Research

This chapter discusses some fundamentals about plastic electronics. The two materials classes and devices investigated in this thesis: organic semiconductors and conductors along with organic transistors and the transparent electrode were also introduced. The importance of organic semiconductor growth and nucleation for transistors was highlighted. Also, the incredible properties of high performance materials like CNTs and graphene for carbon based electronics was discussed. Improving the performance of both transistors and conductors relies on understanding, controlling, and engineering organic semiconductor energy levels, nucleation and growth, which is the focus of the research presented in the remainder of this thesis.

## References

1. Dimitrakopoulos CD, Malenfant PRL (2002) Organic thin film transistors for large area electronics. Adv Mater 14:99
2. Bao Z, Locklin J (2007) Organic field effect transistors (ed. Group, C. P. T. a. F.)
3. Virkar A, Ling MM, Locklin J, Bao Z (2008) Oligothiophene based organic semiconductors with cross-linkable benzophenone moieties. Synth Metals 158:958–963
4. Roberts ME, Sokolov AN, Bao ZN (2009) Material and device considerations for organic thin-film transistor sensors. J Mater Chem 19:3351–3363
5. Bettinger CJ, Bao ZA (2010) Organic thin-film transistors fabricated on resorbable biomaterial substrates. Adv Mater 22:651
6. Ruiz R et al (2004) Pentacene thin film growth. Chem Mater 16:4497–4508
7. Forrest SR (2004) The path to ubiquitous and low-cost organic electronic appliances on plastic. Nature 428:911–918
8. Dimitrakopoulos CD, Mascaro DJ (2001) Organic thin-film transistors: a review of recent advances. Ibm J Res Dev 45:11–27
9. Newman CR et al (2004) Introduction to organic thin film transistors and design of n-channel organic semiconductors. Chem Mater 16:4436–4451
10. Briseno AL et al (2006) Patterning organic single-crystal transistor arrays. Nature 444:913–917
11. Chang PC, Molesa SE, Murphy AR, Frechet JMJ, Subramanian V (2006) Inkjetted crystalline single monolayer oligothiophene OTFTs. Ieee Trans Electron Devices 53:594–600
12. Chua LL et al (2005) General observation of n-type field-effect behaviour in organic semiconductors. Nature 434:194–199
13. Salleo A, Chabinyc ML, Yang MS, Street RA (2002) Polymer thin-film transistors with chemically modified dielectric interfaces. Appl Phys Lett 81:4383–4385
14. Kelley TW et al (2004) Recent progress in organic electronics: materials, devices, and processes. Chem Mater 16:4413–4422
15. Park YD, Lim JA, Lee HS, Cho K (2007) Interface engineering in organic transistors. Mater Today 10:46–54
16. Facchetti A, Yoon MH, Marks TJ (2005) Gate dielectrics for organic field-effect transistors: new opportunities for organic electronics. Adv Mater 17:1705–1725
17. Roberts ME, Mannsfeld SCB, Stoltenberg RM, Bao ZN (2009) Flexible, plastic transistor-based chemical sensors. Org Electron 10:377–383
18. Eder F et al (2004) Organic electronics on paper. Appl Phys Lett 84:2673–2675

19. Dinelli F et al (2004) Spatially correlated charge transport in organic thin film transistors. Phys Rev Lett 92 (116802):1–4
20. Dodabalapur A, Torsi L, Katz HE (1995) Organic Transistors - 2-dimensional transport and improved electrical characteristics. Science 268:270–271
21. Gundlach DJ, Lin YY, Jackson TN, Nelson SF, Schlom DG (1997) Pentacene organic thin-film transistors - Molecular ordering and mobility. Ieee Electron Device Lett 18:87–89
22. Kelley TW, Muyres DV, Baude PF, Smith TP, Jones TD (2003) High performance organic thin film transistors. In: Organic and Polymeric Materials and Devices. Symposium (Mater Res Soc Symposium Proceedings vol 771), 169–79|xiii + 409
23. Lee J et al (2009) Ion gel-gated polymer thin-film transistors: operating mechanism and characterization of gate dielectric capacitance, switching speed, and stability. J Phys Chem C 113:8972–8981
24. Veres J, Ogier S, Lloyd G, De Leeuw D (2004) Gate insulators in organic field-effect transistors. Chem Mater 16:4543–4555
25. Kim C, Facchetti A, Marks TJ (2007) Polymer gate dielectric surface viscoelasticity modulates pentacene transistor performance. Science 318:76–80
26. Knipp D, Street RA, Volkel A, Ho J (2003) Pentacene thin film transistors on inorganic dielectrics: Morphology, structural properties, and electronic transport. J Appl Phys 93:347–355
27. Li XC et al (1998) A highly pi-stacked organic semiconductor for thin film transistors based on fused thiophenes. J Am Chem Soc 120:2206–2207
28. Mannsfeld SCB et al (2007) Selective nucleation of organic single crystals from vapor phase on nanoscopically rough surfaces. Adv Function Mater 17:3545–3553
29. Panzer MJ, Frisbie CD (2005) Polymer electrolyte gate dielectric reveals finite windows of high conductivity in organic thin film transistors at high charge carrier densities. J Am Chem Soc 127:6960–6961
30. Puigdollers J et al (2004) Pentacene thin-film transistors with polymeric gate dielectric. Org Electron 5:67–71
31. Yoon MH, Facchetti A, Marks TJ (2005) Sigma-pi molecular dielectric multilayers for low-voltage organic thin-film transistors. In: Proceedings of the National Academy of Sciences of the United States of America, vol 102, pp 4678–4682
32. Brinkmann M et al (2003) Orienting tetracene and pentacene thin films onto friction-transferred poly(tetrafluoroethylene) substrate. J Phys Chem B 107:10531–10539
33. Fritz SE, Kelley TW, Frisbie CD (2005) Effect of dielectric roughness on performance of pentacene TFTs and restoration of performance with a polymeric smoothing layer. J Phys Chem B 109:10574–10577
34. Heringdorf F, Reuter MC, Tromp RM (2001) Growth dynamics of pentacene thin films. Nature 412:517–520
35. Ito Y et al (2009) Crystalline ultrasmooth self-assembled monolayers of alkylsilanes for organic field-effect transistors. J Am Chem Soc 131:9396–9404
36. Markov I (2003) Crystal growth for beginners: fundamentals of nucleation, crystal growth and epitaxy. In: Scientific W (ed), 2nd edn, New Jersey
37. Verlaak S, Steudel S, Heremans P, Janssen D, Deleuze MS (2003) Nucleation of organic semiconductors on inert substrates. Phys Rev B 68
38. Xue HW, Moyle AM, Magee N, Harrington JY, Lamb D (2005) Experimental studies of droplet evaporation kinetics: Validation of models for binary and ternary aqueous solutions. J Atmospheric Sci 62:4310–4326
39. Ohring M (2001) The material science of thin films. In: Press A (ed), 2nd edn, Orlando
40. Venables JA, Spiller GDT, Hanbucken M (1984) Nucleation and growth of thin-films. Rep Prog Phys 47:399–459
41. Northrup JE, Tiago ML, Louie SG (2002) Surface energetics and growth of pentacene. Phys Rev B (Condensed matter and materials physics) 66: 121404-1–121404-14
42. Virkar A et al (2009) The role of OTS density on Pentacene and C-60 nucleation, thin film growth, and transistor performance. Adv Funct Mater 19:1962–1970

43. Schreiber F (2004) Organic molecular beam deposition: Growth studies beyond the first monolayer. Phys Status Solidi Appl Res 201:1037–1054
44. Amar JG, Family F (1995) Critical cluster-size - island morphology and size distribution in submonolayer epitaxial-growth. Phys Rev Lett 74:2066–2069
45. Brinkmann M, Pratontep S, Contal C (2006) Correlated and non-correlated growth kinetics of pentacene in the sub-monolayer regime. Surf Sci 600:4712–4716
46. Frankl DR, Venables JA (1970) Nucleation on substrates from vapour phase. Adv Phys 19:409
47. Pratontep S, Brinkmann M, Nuesch F, Zuppiroli L (2004) Correlated growth in ultrathin pentacene films on silicon oxide: effect of deposition rate. Phys Rev B 69:165201–165208
48. Pratontep S, Brinkmann M, Nuesch F, Zuppiroli L (2004) Nucleation and growth of ultrathin pentacene films on silicon dioxide: effect of deposition rate and substrate temperature. Synth Metals 146:387–391
49. Pratontep S, Nuesch F, Zuppiroli L, Brinkmann M (2005) Comparison between nucleation of pentacene monolayer islands on polymeric and inorganic substrates. Phys Rev B 72: 085211–0852216
50. Ruiz R et al (2003) Dynamic scaling, island size distribution, and morphology in the aggregation regime of submonolayer pentacene films. Phys Rev Lett 91(136102):1–4
51. Venables JA et al (1973) Rate equation approaches to thin-film nucleation kinetics. Philos Mag 27:697–738
52. Zinsmeis G (1969) Theory of thin film condensation.C. aggregate size distribution in island films. Thin Solid Films 4:363
53. Zinsmeister G (1966) A contribution to Frenkel's theory of condensation. Vacuum 16:529
54. Yang HC et al (2005) Conducting AFM and 2D GIXD studies on pentacene thin films. J Am Chem Soc 127:11542–11543
55. Steudel S, Janssen D, Verlaak S, Genoe J, Heremans P (2004) Patterned growth of pentacene. Appl Phys Lett 85:5550–5552
56. Steudel S et al (2004) Influence of the dielectric roughness on the performance of pentacene transistors. Appl Phys Lett 85:4400–4402
57. Choudhary D, Clancy P, Bowler DR (2005) Adsorption of pentacene on a silicon surface. Surf Sci 578:20–26
58. Choudhary D, Clancy P, Shetty R, Escobedo F (2006) A computational study of the sub-monolayer growth of pentacene. Adv Funct Mater 16:1768–1775
59. Fritz SE, Martin SM, Frisbie CD, Ward MD, Toney MF (2004) Structural characterization of a pentacene monolayer on an amorphous SiO2 substrate with grazing incidence X-ray diffraction. J Am Chem Soc 126:4084–4085
60. Mannsfeld SCB, Virkar A, Reese C, Toney MF, Bao ZN (2009) Precise structure of pentacene monolayers on amorphous silicon oxide and relation to charge transport. Adv Mater 21:2294
61. Ruiz R et al (2004) Structure of pentacene thin films. Appl Phys Lett 85:4926–4928
62. Chen ZX, Ikeda S, Saiki K (2006) Sexithiophene films on cleaved KBr(100) towards well-ordered semiconducting films. Mater Sci Eng B-Solid State Mater Adv Technol 133:195–199
63. Era M, Tsutsui T, Saito S (1995) Polarized electroluminescence from oriented p-sexiphenyl vacuum-deposited film. Appl Phys Lett 67:2436–2438
64. Halik M et al (2003) Relationship between molecular structure and electrical performance of oligothiophene organic thin film transistors. Adv Mater 15:917
65. Horowitz G, Hajlaoui ME (2000) Mobility in polycrystalline oligothiophene field-effect transistors dependent on grain size. Adv Mater 12:1046–1050
66. Li RJ et al (2009) Micrometer- and nanometer-sized, single-crystalline ribbons of a cyclic triphenylamine dimer and their application in organic transistors. Adv Mater 21:1605
67. Ling MM et al (2007) Air-stable n-channel organic semiconductors based on perylene diimide derivatives without strong electron withdrawing groups. Adv Mater 19:1123–1127
68. Liu SH et al (2009) Patterning of alpha-Sexithiophene single crystals with precisely controlled sizes and shapes. Chem Mater 21:15–17

69. Liu YL et al (2006) Controlling the growth of single crystalline nanoribbons of copper tetracyanoquinodimethane for the fabrication of devices and device arrays. J Am Chem Soc 128:12917–12922
70. Loi MA et al (2005) Supramolecular organization in ultra-thin films of alpha-sexithiophene on silicon dioxide. Nature Mater 4:81–85
71. Dimitrakopoulos CD, Brown AR, Pomp A (1996) Molecular beam deposited thin films of pentacene for organic field effect transistor applications. J Appl Phys 80:2501–2508
72. Cantrell R, Clancy P (2008) A computational study of surface diffusion of C-60 on pentacene. Surf Sci 602:3499–3505
73. Engstrom JR, Goose JE, Killampalli A, Clancy P (2009) Molecular-scale events in hyperthermal deposition of organic semiconductors implicated from experiment and molecular simulation. J Phys Chem C 113:6068–6073
74. Forrest SR (1997) Ultrathin organic films grown by organic molecular beam deposition and related techniques. Chem Rev 97:1793–1896
75. Laquindanum JG, Katz HE, Dodabalapur A, Lovinger AJ (1996) n-channel organic transistor materials based on naphthalene frameworks. J Am Chem Soc 118:11331–11332
76. Shtein M, Mapel J, Benziger JB, Forrest SR (2002) Effects of film morphology and gate dielectric surface preparation on the electrical characteristics of organic-vapor-phase-deposited pentacene thin-film transistors. Appl Phys Lett 81:268–270
77. Sirringhaus H et al (1999) Two-dimensional charge transport in self-organized, high-mobility conjugated polymers. Nature 401:685–688
78. Lin YY, Gundlach DJ, Nelson SF, Jackson TN (1997) Stacked pentacene layer organic thin-film transistors with improved characteristics. Ieee Electron Device Lett 18:606–608
79. Lin YY, Gundlach DJ, Nelson SF, Jackson TN (1997) Pentacene-based organic thin-film transistors. Ieee Trans Electron Devices 44:1325–1331
80. Liu SH et al (2007) Selective crystallization of organic semiconductors on patterned templates of carbon nanotubes. Adv Funct Mater 17:2891–2896
81. Knipp D, Street RA, Volkel AR (2003) Morphology and electronic transport of polycrys talline pentacene thin-film transistors. Appl Phys Lett 82:3907–3909
82. Locklin J, Bao ZN (2006) Effect of morphology on organic thin film transistor sensors. Anal Bioanal Chem 384:336–342
83. Schwoebe.Rl & Shipsey, E. J. Step Motion on Crystal Surfaces. Journal of Applied Physics 37, 3682-& (1966)
84. Shtein M et al (2002) Organic VPD shows promise for OLED volume production (vol 45, pg 131, 2002). Solid State Technol 45:18
85. Liu SH, Wang WCM, Briseno AL, Mannsfeld SCE, Bao ZN (2009) Controlled deposition of crystalline organic semiconductors for field-effect-transistor applications. Adv Mater 21:1217–1232
86. Geim AK, Novoselov KS (2007) The rise of graphene. Nature Mater 6:183–191
87. Castro Neto AH, Guinea F, Peres NMR, Novoselov KS, Geim AK (2009) The electronic properties of graphene. Rev Modern Phys 81: 109–162
88. Ebbesen TW et al (1996) Electrical conductivity of individual carbon nanotubes. Nature 382:54–56
89. Geng HZ et al (2008) Doping and de-doping of carbon nanotube transparent conducting films by dispersant and chemical treatment. J Mater Chem 18:1261–1266
90. Hellstrom SL, Lee HW, Bao ZN (2009) Polymer-assisted direct deposition of uniform carbon nanotube bundle networks for high performance transparent electrodes. Acs Nano 3:1423–1430

# Chapter 2
# Organic Semiconductor Growth and Transistor Performance as a Function of the Density of the Octadecylsilane Dielectric Modification Layer

## 2.1 Introduction

The central research focus in organic electronics has been improvement of charge transport in organic thin film transistors (OTFTs), the building blocks of organic circuits [1, 2]. Charge transport takes place primarily within the first few monolayers of semiconductor at the dielectric/semiconductor interface; therefore device performance is dominated by the properties of these interfacial layers [3, 4]. As aforementioned, grain boundaries between crystalline domains affect device performance, acting as barriers to charge transport [1, 5, 6]. Consequently, highly crystalline films deposited via a layer-by-layer 2D growth mode are desirable due to less energetically severe in-plane boundaries.

Octadecylsilanes (OTS), specifically octadecyltrimethoxysilane (OTMS) and octadecyltrichlorosilane (OTCS), have been widely used to modify the $SiO_2$ dielectric surface and have resulted in dramatic improvement of the field-effect mobility for a variety of semiconductors [4, 7–10]. In fact OTS modified $SiO_2$ is the most common surface for analyzing organic semiconductors [1, 2]. The hydrophobic nature of OTS is thought to passivate the $SiO_2$ surface, increase semiconductor crystal quality, reduce interfacial trap states and in some cases planarize the surface [5, 11, 12]. Previous reports have attributed the improvement of mobility on OTS-treated surfaces to a drastic reduction in surface energy by hydrophobic surface modification [13, 14]. However, a wide range of charge-carrier mobilities have been reported by a number of research groups for the same organic semiconductor on OTS, suggesting the need to understand the nature of the underlying OTS layer. For example, the reported mobilities of pentacene, the most widely studied semiconductor, OTFTs with OTS-modified $SiO_2$ ranges from 0.03 to greater than 2.0 $cm^2 V^{-1} s^{-1}$ [1, 15]. This reflects the difference in a technologically useless semiconducting thin film (0.03 $cm^2 V^{-1} s^{-1}$) to semi-conducting thin film whose mobility (2.0 $cm^2 V^{-1} s^{-1}$) exceeds that of amorphous silicon.

A. Virkar, *Investigating the Nucleation, Growth, and Energy Levels of Organic Semiconductors for High Performance Plastic Electronics*, Springer Theses, DOI: 10.1007/978-1-4419-9704-3_2, © Springer Science+Business Media, LLC 2012

Previous work has shown that both inorganic and organic materials growth is highly sensitive to the chemical nature, packing, and defects of the underlying self-assembled monolayer (SAM) [16]. Recently, Cho and co-workers studied the dependence of pentacene TFTs on OTS order. They found improved pentacene TFT performance when the OTS layer was prepared at lower temperatures so that the monolayer was ordered [17]. However, conclusive and quantitative reasons for the differences in growth mode, and number of transistors tested were not discussed. In fact, prior to the research presented in this thesis, there has been almost no quantitative experimental research into nucleation and thin film growth of pentacene on organic surfaces (like OTS) even though they are more useful and common than inorganic ones like bare (unmodified) $SiO_2$. This is primarily because forming a reproducible OTS layer has been challenging. Cho and co-workers suggested that a lower nucleation density is observed on ordered OTS due to a greater pentacene diffusivity (see Chap. 3). In an earlier report, Cho and co-workers studied the chain length dependence of alkyl-silane treated $SiO_2$ on pentacene TFTs. In that report they found that pentacene diffusivity was the largest and nucleation density was the lowest on the most disordered shorter chained alkylsilanes [17]. However, in both of these reports the role of the surface on heterogeneous nucleation was not considered, which is the focus of the next two chapters, and is a major factor governing OTFT performance. The incongruence in Cho and co-workers' results stems from the fact that both growth mode and nucleation density are highly sensitive to the surface. Furthermore, models which have been developed for inorganic nucleation and thin film growth often fail to accurately represent organic thin film growth due to the differences in molecular symmetry and bonding.

In addition, some semiconductors only show marginally improved performance on any OTS treated surface compared to $SiO_2$ despite the large difference in surface energy [5]. This suggests, as Markov and others have theorized, that each combination of molecule and substrate must be considered unique [18, 19]. The most relevant factor affecting growth, and consequently charge carrier mobility, is the specific energetic interaction between the semiconductor and the surface [18, 19]. This interaction is influenced considerably by the phase of the underlying OTS monolayer, even though all the OTS phases studied have identical chemical compositions and similar surface energies and roughness.

## 2.2 Fabrication and Characterization of Octadecylsilane Monolayers

In the study presented in this chapter, the effect of density and degree of ordering of OTS monolayers on the performance of two of the most widely studied and highest mobility thin film organic semiconductors: pentacene (p-channel) and $C_{60}$ (n-channel) was investigated [1, 4, 5, 11, 20–24]. The Langmuir–Blodgett (LB)

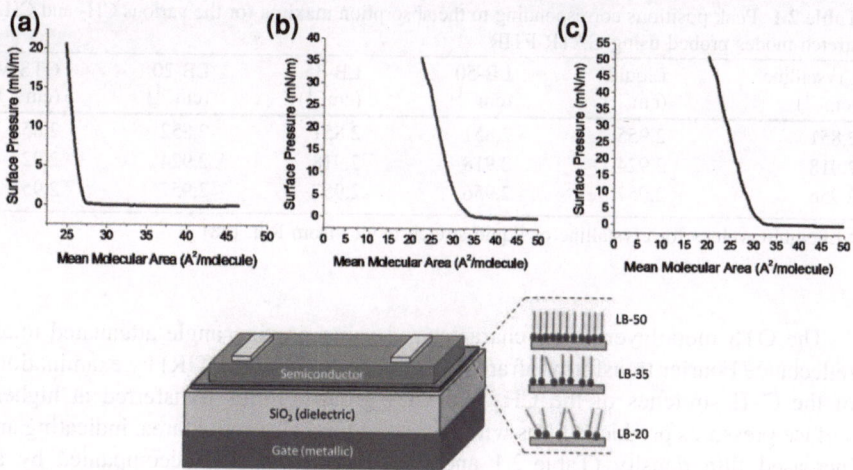

**Fig. 2.1** Representative surface pressure-area isotherms for **a** LB-20 mN m$^{-1}$. (LB-20), **b** LB-35 mN m$^{-1}$ (LB-35), and **c** LB-50 mN m$^{-1}$ (LB-50). The typical film collapse pressure was 55 mN m$^{-1}$. Below is a schematic highlighting that the only variable between the transistors studied was the underlying order of the OTS SAM

technique was employed to systematically vary the organization and density of the OTS monolayers. In this well-known ultrathin-film deposition technique, amphiphilic OTS molecules are compressed by applying a lateral pressure to the monolayer film at the air–water interface [25]. Under increasing applied lateral pressure, the film undergoes a transition from a 2D gas to a 2D liquid and finally to an ordered 2D solid [25]. From the Langmuir isotherms obtained in our study, the OTS film collapses at a surface pressure of ~55 mN m$^{-1}$ and there appears to be a phase transition leading to the most ordered phase at a surface pressure of ~40mN m$^{-1}$ (Fig. 2.1). This phase change from one condensed (2D solid) phase to another condensed phase was also observed by Duran and co-workers [26]. Accordingly, surface pressures of 20, 35, and 50 mN m$^{-1}$ were chosen to study OTS films of different degrees of order (designated as LB-20, LB-35 and LB-50). Moreover, by studying LB-35 and LB-50 we were able to probe two distinct condensed phases. During fabrication of the Langmuir films, OTMS molecules were hydrolyzed and partially polymerized on the trough at pH of 3 [26]. The polymerized monolayer was Blodgett-transferred to the thermally grown silicon oxide (300 nm) surface on a heavily doped silicon wafer, which is used as the gate electrode. For comparison, the commonly used OTMS-V and OTCS-V vapor deposited films were also prepared. These vapor deposited films are considerably less dense than the LB films, and are more susceptible to inconsistencies in film properties since they cannot be fabricated in a controlled way like LB films [4, 27]. Transistors were completed in top contact geometry with gold source and drain electrodes (see Experimental). The channel length was 50 μm and the width was 1,000 μm. Thus the only variable probed was the order and density of the underlying OTS.

**Table 2.1** Peak positions corresponding to the absorption maxima for the various $CH_2$ and $CH_3$ stretch modes probed using GATR-FTIR

| Crystalline[a] $(cm^{-1})$ | Liquid[a] $(cm^{-1})$ | LB-50 $(cm^{-1})$ | LB-35 $(cm^{-1})$ | LB-20 $(cm^{-1})$ | OTS-V $(cm^{-1})$ |
|---|---|---|---|---|---|
| 2,851 | 2,955 | 2,851 | 2,851 | 2,852 | 2,852 |
| 2,918 | 2,924 | 2,918 | 2,918 | 2,924 | 2,924 |
| 2,956 | 2,957 | 2,956 | 2,957 | 2,957 | 2,957 |

[a] Literature values for crystalline or liquid stretch modes from Ref. [28]

The OTS monolayers were characterized using grazing angle attenuated total reflectance Fourier transform infrared spectroscopy (GATR-FTIR) by examination of the C–H stretches of the $CH_2$ and $CH_3$ groups. Films transferred at higher surface pressures produced films with increased total absorption area, indicating an increased film density (Table 2.1 and Fig. 2.3a). This was accompanied by a characteristic shift of aliphatic vibrational stretching modes to lower wavenumbers (from 2,924 to 2,918 $cm^{-1}$ for the asymmetric C–H stretch and from 2,855 to 2,851 $cm^{-1}$ for the symmetric C–H stretch), indicating a transition from a liquid-like (disordered) to crystalline layer [28]. The magnitude of the observed peak shifts is similar to those reported in literature. The results are summarized in Table 2.1.

The OTS monolayers were further characterized using grazing incidence X-ray diffraction (GIXD). GIXD can probe in-plane order, and it is the ideal technique to study the crystalline order of a monolayer. Only the highly dense LB-50 mono-layer gave rise to a Bragg rod in the GIXD spectrum (Figs. 2.2 and 2.6d). This indicates that indeed only the LB-50 monolayer has crystalline order. The cal-culated 4.2 Å hexagonal lattice constant for the LB-50 OTS is consistent with previous reports for crystalline OTS [29]. The LB-OTS peak does show some mild arching (Fig. 2.2). It is likely this tilt is due to the defects in the OTS film which arise from some ordered domains being non-coplanar with the majority of the film (Fig. 2.2c). Such defects can arise in LB films since the films are prepared by applying lateral compression.

The density and order of the LB films as a function of surface pressure were further characterized by ellipsometry, static water contact angle, and high resolu-tion AFM (Table 2.2). As the monolayers were compressed, the film thickness and water contact angle increased, again indicating a denser monolayer. Using high resolution atomic force microcopy, it was determined that the LB films also showed larger overall domain size, roughly 30% larger than the vapor phase deposited films. The root mean square (RMS) roughness was similar for the LB films and the vapor deposited OTS-V film, confirming that lower mobilities observed for OTS-V treated OTFTs are not due to surface roughness effects [12, 13, 30]. The key film characteristics (C–H stretching mode frequencies, contact angle and surface roughness) for all the OTS films remained unchanged after heating to 200 °C under argon, indicating the films are stable and do not undergo thermal phase changes in the temperature ranges used for semiconductor deposition.

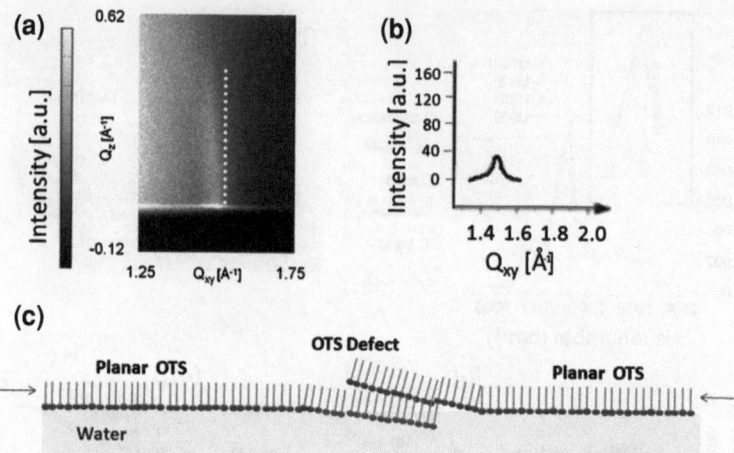

**Fig. 2.2 a** GIXD spectrum of LB-50 OTS monolayer. The white dotted line is drawn perpendicular to highlight the mild arching in the OTS peak. **b** Line profile of OTS LB-50 spectrum. **c** Schematic of OTS film on Langmuir trough showing potential defects which can arise during compression. The arrows indicate the direction of compression

**Table 2.2** Properties of OTS monolayers studied

| Sample | Mean molecular area ($\mathring{A}^2$ molecular$^{-1}$) | Height (nm) | Contact angle (deg) | RMS roughness (nm) |
|--------|------------------|-------------|---------------------|---------------------|
| LB-20 | 24.8 | 2.0 | 101.7 | 0.2 |
| LB-35 | 22.5 | 2.1 | 102.6 | 0.3 |
| LB-50 | 20.1 | 2.1 | 104.1 | 0.2 |
| OTS-V | 28.7[a] | 1.9 | 98.3 | 0.2 |

[a] The MMA values were estimated by calculating the area under absorption peaks from GATR-FTIR spectra for the OTS V surface and comparing them to the area under the absorption peaks for the LB films with known MMA values

## 2.3 Effects of OTS Density on Pentacene and C60 Transistor Performance

The pentacene charge-carrier mobility ($\mu$) extracted from saturation transfer characteristics is plotted as a function of the density [inverse mean molecular area (MMA)] of the OTS monolayer in Fig. 2.3c. The MMA is calculated based on the total number of molecules deposited on the Langmuir trough, and the total area occupied by the Langmuir film at the corresponding surface pressure. The average pentacene hole mobility measured for 50 devices on each type of OTS increases with increasing OTS density from 0.4 cm$^2$ V$^{-1}$ s$^{-1}$ for the least compressed LB film (LB-20) to 1.9 cm$^2$ V$^{-1}$ s$^{-1}$ on the most compressed (highest order) LB film (LB-50) despite only small differences in surface energy and roughness. This suggests that the density of OTS is a critical factor effecting performance. The mobility

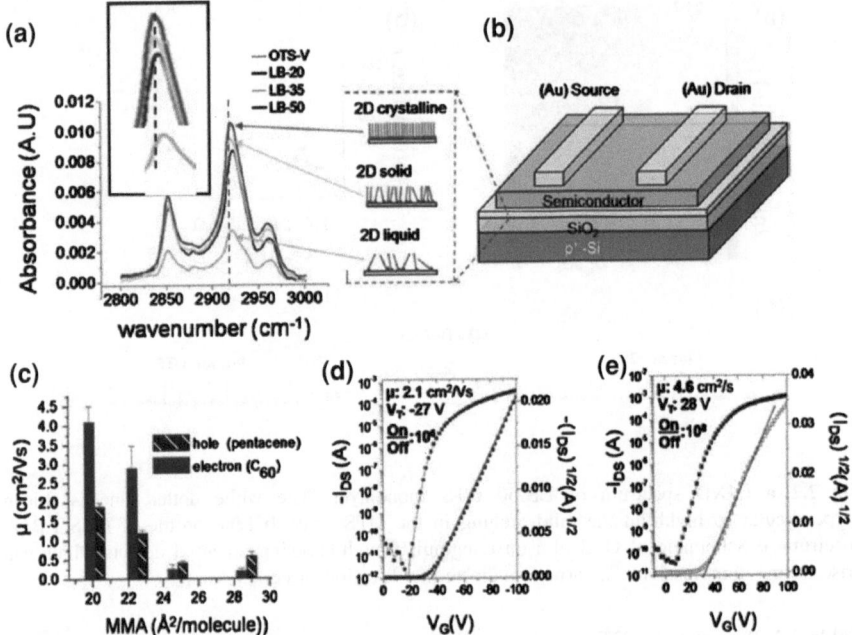

**Fig. 2.3 a, b** GATR–FTIR spectrum for the OTS films with differing 2D phases, and densities quantified by the mean molecular area MMA extracted from the Langmuir trough. **b** Schematic of pentacene OTFT with underlying OTS monolayers in different 2D phases. **c** The average (over 50 devices per OTS treatment) saturation mobility $\mu$ ($cm^2\ V^{-1}\ s^{-1}$) of 45 nm pentacene OTFTs measured in ambient and average (over 10 devices per OTS treatment) saturation mobility $\mu$ ($cm^2\ V^{-1}\ s^{-1}$) of 45 nm $C_{60}$ OTFTs tested in a $N_2$ glovebox. The source–drain voltage was fixed at $-100$ V for p-channel transistors (pentacene), and 100 V for n-channel transistors ($C_{60}$). **d** Typical I–V transfer curves for pentacene TFTs with LB-50 OTS treatment. The mobility, threshold voltage and on/off ratios are provided as insets. **e** Typical I–V transfer curves for $C_{60}$ TFTs with LB-50 OTS treatment. The mobility, threshold voltage VT and on/off ratios are provided as insets

on the less dense OTS-V (MMA $= 28.7$ Å$^2$ molecule$^{-1}$) is comparable to the lowest compressed LB film (20 m Nm$^{-1}$, MMA $= 24.8$ Å$^2$ molecule$^{-1}$) substrate despite lower film density. See Tables 2.3 and 2.4 for summary of average electrical characteristics. $C_{60}$ OTFTs showed an electron mobility as high as 5.2 cm$^2$ V$^{-1}$s$^{-1}$ (measured in a nitrogen glovebox) on the LB-50 film and followed the same trend as pentacene with mobility decreasing with decreasing OTS order. This $C_{60}$ mobility on the dense OTS is amongst the highest reported in the literature [4, 24].

The pentacene TFTs were fabricated on four different days during four different depositions, and the $C_{60}$ TFTs were deposited on two different days. For each respective set of devices the performance was very similar on LB-50 regardless of the deposition and the performance was consistently higher than the OTS-V treated TFTs. Both types of transistors showed negligible change in the threshold voltage on the various OTS surfaces, the on/off ratio was consistently above $10^5$–$10^6$ and the gate current was consistently several orders of magnitude lower than the drain current (see Figs. 2.4 and 2.5). For 50 μm channel length TFTs the contact resistances, are typically negligible compared to channel resistances; the device performance is dominated by the channel, and thus by the pentacene grains in the channel [31, 32]. From the output characteristics, the linear region of the IV curves show no non-linearity indicating the absence of contact issues. In order to test for the contribution of contact and channel resistances, several pentacene OTFTs with different channel lengths were tested and contact resistance was extracted. The contact resistances were nearly identical (within 3.0%) for TFTs fabricated on different OTS monolayers and were much smaller than the channel resistances indicating that indeed the channel effects dictate performance. (Fig. 2.5).

## 2.4 Organic Semiconductor Thin Film Analysis: Grazing Incidence X-ray Diffraction and Atomic Force Microscopy

In order to determine the effect of OTS monolayer density on organic semiconductor nucleation and growth at the semiconductor/dielectric interface, pentacene thin films (nominally 3 nm monitored by quartz microbalance during deposition) were deposited onto the OTS films under identical conditions as those used for OTFT fabrication. Due to the gate field, in the transistor's "on" state, the majority of charge carriers are induced and transported in the first $\sim 5$ nm of semiconductor near the dielectric interface; the packing and morphology of the initially deposited interfacial layer is therefore critical [3, 33]. These were examined using atomic force microscopy (AFM) and grazing incidence X-ray diffraction (GIXD), giving information about the morphology and crystalline order of the pentacene monolayer film directly involved in charge transport. AFM was performed immediately after deposition to ensure that the film did not undergo reorganization. AFM was also performed before and after GIXD experiments to also ensure that films did not change during exposure to X-rays (typically for 30 min).

To determine if the packing of pentacene in the first monolayer is affected by the difference in OTS density, we carried out GIXD of the pentacene monolayers. The characteristic pentacene (11L), (02L) and (12L) in-plane Bragg rods are seen in the GIXD spectra (Fig. 2.6). On the LB-50 film (Fig. 2.6d), an additional broad peak between the (11L) and (02L) pentacene peaks is observed again due to the crystalline nature of the underlying OTS. The lattice constants of pentacene ($a = 5.93$ Å, $b = 7.58$ Å, $\gamma \approx 90°$) extracted from the diffraction peaks were nearly identical regardless of the OTS preparation method and are similar to those reported for pentacene grown on hexamethyldisilazane (HMDS) or OTS [34, 35]. The pentacene GIXD spectra (position of peaks in $Q_{xy}$ and $Q_z$) are also similar on all the OTS surfaces. This indicates that the difference in mobility on different OTS surfaces is not due to different pentacene packing motifs. It is also interesting

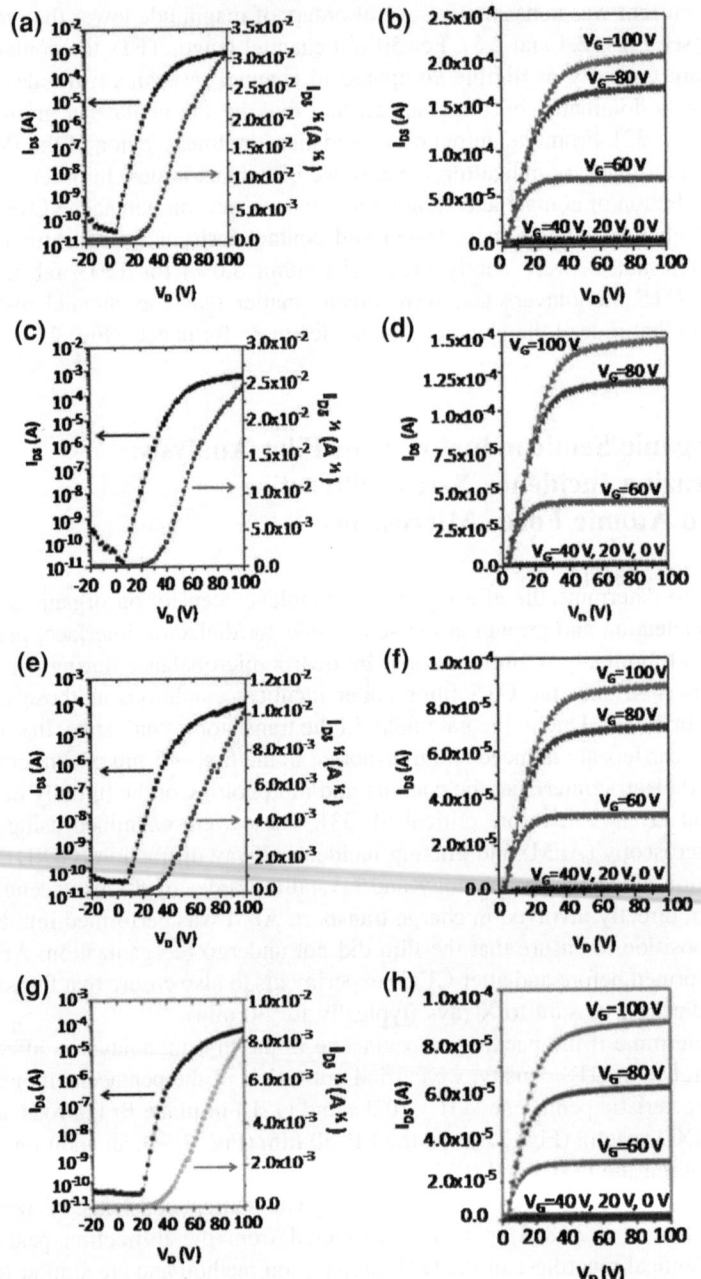

**Fig. 2.4** Transfer and output curves for $C_{60}$ 45 nm TFT on different OTS layer. **a** transfer curve LB-50, **b** output curves LB-50, **c** transfer curves LB-35, **d** output curves LB-35, **e** transfer curves LB-20, **f** output curves LB-20, **g** transfer curve OTS-V, **h** output curves OTS-V

**Fig. 2.5  a** Resistance versus channel length for pentacene TFTs with OTS-V or LB-50 dielectric modification layers. The equation of the line of best fit and the corresponding $R^2$ value are provided on the graph. The equation is written in a $y = mx + b$ format, where $m$ is the slope relating the resistance ($y$) to channel length ($x$), and extracting to $x = 0$ the contact resistance can be evaluated as b/2. Thus the extracted two point contact resistance for typical LB-50 TFTs is 2.67 k$\Omega$ and for OTS-V TFT is 2.75 k$\Omega$. The difference in contact resistance is <3%. **b** Typical gate and drain currents for pentacene TFTs with LB-50 dielectric modification layers. Typical gate currents are orders of magnitude smaller than drain currents

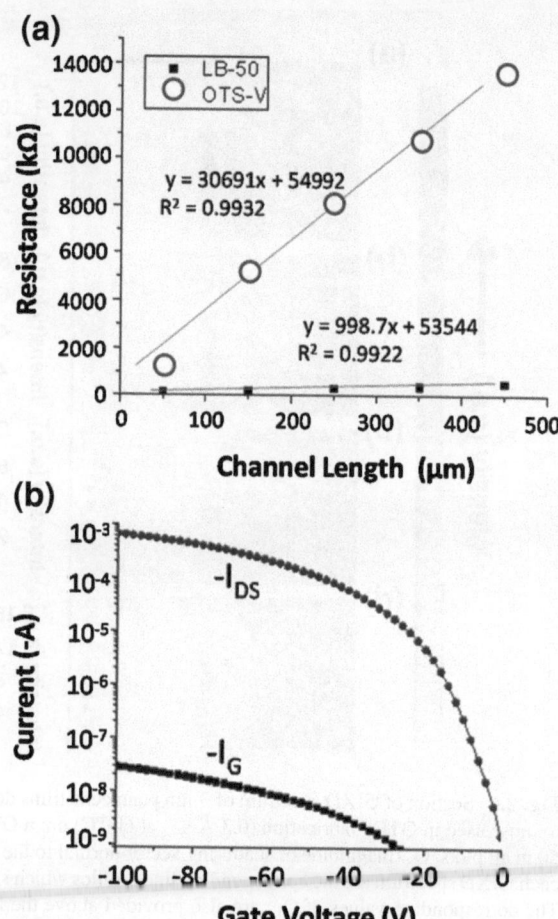

to note there is an additional diffraction peak at $Q_{xy} = 1.6 \text{ Å}^{-1}$ on OTS-V (Fig. 2.6a) which corresponds to a portion of the film exhibiting the bulk pentacene phase. This 3D growth on OTS-V is further asserted by the AFM results discussed below. The full-width at half max (FWHM) of the diffraction peaks can be used to gauge the crystalline quality of the pentacene on various OTS surfaces. However, for all the films studied, the FWHM was resolution limited (domain size >10 nm) (see Experimental 2.7).

AFM of the nominally 3 nm pentacene films (Fig. 2.7) did, however, show a clear trend between thin film morphology and charge carrier mobility. The growth mode on the highly ordered LB films showed the more desirable 2D layer-by-layer (Frank-van der Merwe type) growth (Fig. 2.7d) as compared to the less favorable 3D (Volmer-Weber type) growth on OTS-V (Fig. 2.7a) which leads to many island severe grain boundaries after coalescence [1, 18, 19]. The growth mode on OTS-V is purely 3D island-type; no complete monolayer forms within the first 3 nm of

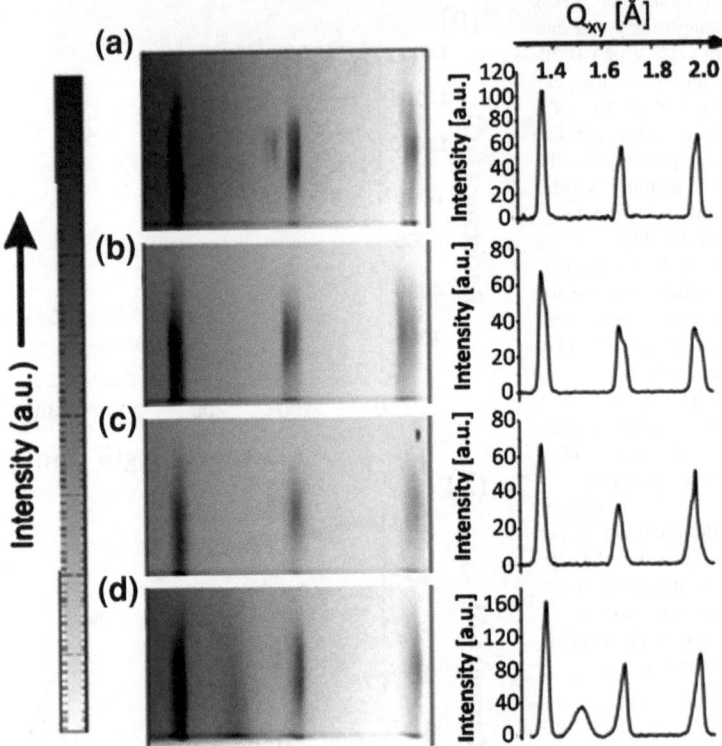

**Fig. 2.6** Section of GIXD spectrum of 3 nm pentacene films deposited under identical conditions to those used in OTFT fabrication (0.3 Å s$^{-1}$ at 60 °C) on: **a** OTS-V, **b** on LB-20 **c** LB-35, **d** LB-50 in all plots, $Q_z$ (magnitude of scattering vector normal to the surface) is vertical. To the right of each GIXD spectrum are the corresponding line profiles which show the integrated peaks along $Q_z$. The corresponding values of $Q_{xy}$ are also provided above the line profiles

nominal film thickness before additional layer growth, though the individual crystalline islands are large and terraced. The island size is considerably larger on OTC-V than on the LB films, which are composed of smaller connected islands. This is a critical observation: the growth mode has an enormous effect on the mobility even when the TFTs are fabricated using identical materials, at the same deposition conditions.

## 2.5 Organic Semiconductor Crystallization and 2D Versus 3D Growth Mode on OTS of Varying Order

Due to its paramount importance on the conductivity of thin films of pentacene, the 2D versus 3D growth mode for pentacene is described below and the strength of pentacene OTS interactions are estimated for the different surfaces. Moreover, the

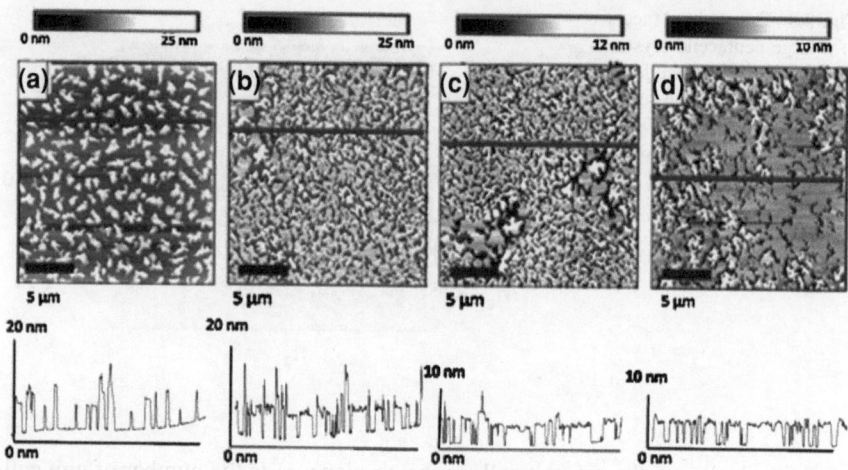

**Fig. 2.7** Nominally 3 nm thin film of pentacene deposited under identical conditions as used in TFT fabrication (0.3 Å s$^{-1}$ at 60 °C) on **a** OTS-V **b** OTS-20, **c** LB-35, **d** LB-50. The line corresponds to the profile shown directly below each AFM image. The growth mode of pentacene tends to be more and more 2D as the OTS density increases

interaction energies needed to drive 2D growth will be calculated, and related to the density of the underlying OTS.

It is important to note that the majority of this analysis considers a simple Kossel crystal (composed of cubes) since more complex crystals are often analytically impossible to analyze [18, 19]. The physics remain the same, but growth models are constructed from this simplified view so analytical expressions can be used and so that predictions of interaction energies are possible. For pentacene approximation as a Kossel crystal is reasonable since: (1) its molecular shape can be approximated as a rectangle, (2) the first monolayers of pentacene stand nearly upright on the substrate, i.e., ~ zero tilt angle, (3) the molecules are symmetric (if you cut a standing pentacene molecule in the middle lengthwise, or in the center, the two halves are identical). (see Appendix for more the general thermodynamic derivation of the 2D versus 3D nucleation a Kossel crystal).

The formation of a 2D or 3D crystal is related to the chemical potential driving force for nucleation, the various surface energies of the crystal, the interfacial energies, and the molecule substrate interaction energy. The change in free energy for a finite sized pentacene crystal (which has two molecules in its unit cell) is related to the crystal size and the chemical potential by (Eq. 2.1, also see Fig. 2.8) [18, 19]:

$$\Delta G = -n_{ab}n_c 2\Delta\mu + [2n_b n_c \gamma_{100} + 2n_a n_c \gamma_{010} + 2n_d n_c \gamma_{1-10}$$
$$+ 2n_e n_c \gamma_{110} + n_{ab}(\gamma_{110} + \gamma_s)] \qquad (2.1)$$

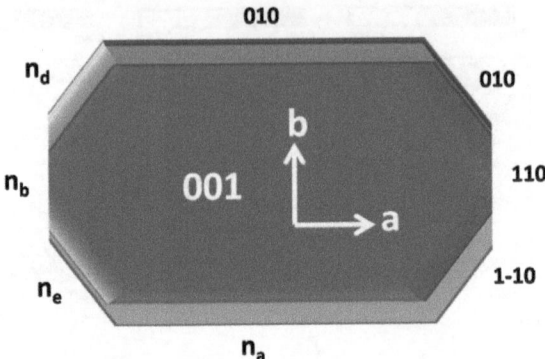

**Fig. 2.8** The various facets of a large pentacene crystal. The number of unit cells corresponding to each surface is also given

where $n_{ab}$ is the number of unit cells in the ab-plane, $n_c$ is the number of unit cells in the c direction (how many unit cells high the crystal is), $\Delta\mu$ is the chemical potential difference between the vapor phase and the crystalline phase (of infinite size), $n_a$, $n_b$, $n_d$, and $n_e$, are the number of unit cells corresponding to the 010, 110, 110, and 1–10 surfaces. The specific surface energy terms (per unit cell) are denoted by $\gamma_i$ ($i$ again referring to the difference surfaces) and $\gamma_s$ is the specific interfacial energy which relates the difference in interlayer interaction energy between layers of pentacene ($\gamma_{001}$) and between pentacene and the substrate ($\gamma_{mol-sub}$). More rigorously defined, it is the difference in energy (per unit cell) at the substrate, between interlayer adhesion energy and the adhesion energy of the surface ($\sigma_i/A_{ab}$ where $A_{ab}$ is the area per unit cell).

The critical cluster size, i.e., the number of unit cells needed to form a stable island, can be approximated by differentiation of (2.1) with respect to $n_a$, $n_b$, $n_d$, $n_c$, and $n_e$ and equating to zero. Each solution is related to $n_c{}^*$ (the critical dimension of how many unit cells high the crystal is), $\Delta\mu$, and the surface/interaction energies. For discussion of 2D versus 3D growth mode $n_c{}^*$ is the most important term. The critical number of how many unit cells tall the crystal is, $n_c$, has the following solutions [19]:

$$n_c^* = \left\{ \left[ \left( \frac{2(\gamma_{001} + \gamma_s)}{2\Delta\mu} \right), 1 \right] \right\}, \tag{2.2}$$

The first solution is for the 3D growth case (any size greater than 1 monolayer in height). The solution "1" of course is for the 2D crystal case. The solutions to the free energy at critical cluster size represent the energetic barrier which is needed to be overcome for nucleation. The solutions for a 3D and 2D crystal are given below [19]:

**Table 2.3** Summary of pentacene TFT data measured in ambient conditions

| Surface treatment | Average ($cm^2 V^{-1} s^{-1}$) | Max ($cm^2 V^{-1} s^{-1}$) | $I_{on}/I_{off}$ | $V_T$ (V) |
|---|---|---|---|---|
| LB-20 | 0.4 (0.05) | 0.6 | $10^6$ | −18 |
| LB-35 | 1.2 (0.08) | 1.4 | $2 \times 10^6$ | −24 |
| LB-50 | 2.1 (0.12) | 2.3 | $2 \times 10^6$ | −20 |
| OTS-V | 0.6 (0.1) | 0.9 | $10^6$ | −19 |

The values are average over $\sim 50$ devices for each OTS substrate treatment

**Table 2.4** Summary of $C_{60}$ OTFT data measured in a nitrogen glovebox

| Surface treatment | Average ($cm^2 V^{-1} s^{-1}$) (SD) | Max ($cm^2 V^{-1} s^{-1}$) | $I_{on}/I_{off}$ | $V_T$ [V] |
|---|---|---|---|---|
| LB-20 | 0.3 (0.1) | 0.4 | $3 \times 10^5$ | 35 |
| LB-35 | 2.9 (0.5) | 3.5 | $4 \times 10^7$ | 34 |
| LB-50 | 4.1 (0.4) | 5.3 | $3 \times 10^7$ | 33 |
| OTS-V | 0.2 (0.1) | 0.3 | $2 \times 10^6$ | 40 |

The values are averaged over ten devices for each OTS substrate treatment

$$\Delta G_{3D}^* = \frac{4(\gamma_{001} + \gamma_s)[(2(\gamma_{010} + \gamma_{100})(\gamma_{110} + \gamma_{1-10})) - (2(\gamma_{010}^2 + \gamma_{100}^2) + \gamma_{110}^2 + \gamma_{1-10}^2)]}{(2\Delta\mu)^2}$$

(2.3)

$$\Delta G_{2D}^* = \frac{2[(\gamma_{010} + \gamma_{100})(\gamma_{110} + \gamma_{1-10})] - [2(\gamma_{010}^2 + \gamma_{100}^2) + \gamma_{110}^2 + \gamma_{1-10}^2]}{2\Delta\mu - (\gamma_{001} + \gamma_s)}$$

(2.4)

Recall if $\Delta G_{3D}^*$ is lower, then 3D nucleation is favored, whereas if $\Delta G_{2D}^*$ is lower in energy than 2D nucleation is favored. From Eq. 2.3, 3D nucleation can occur for all supersaturations $\Delta\mu > 0$. For supersaturated systems, 2D nucleation can occur only once a supersaturation value $\Delta\mu_2$ has been achieved (see Appendix) [18],

$$\Delta\mu_2 = \frac{(\gamma_{001} + \gamma_s)}{2} = \gamma_{\text{interlayer}} - \gamma_{\text{mol-substrate}}$$

(2.5)

since at $\Delta\mu > \Delta\mu_2$ the solution to Eq. 2.5 becomes physically meaningful. As $\Delta\mu$ increases (beyond $\Delta\mu_2$), there is a transition where the barriers of nucleation for 2D and 3D nucleation become identical i.e. $(\Delta G_{2D}^* = \Delta G_{3D}^*)$, and of course this means the 3D nuclei is one monolayer high or simply a 2D crystal! This value of critical supersatution ($\Delta\mu_{cr} = 2\Delta\mu_2$) is twice the transition where 2D nucleation becomes possible [18]. Of course for a given $\Delta\mu$, the differences in the interlayer and molecule–substrate interaction energies will dictate whether 2D or 3D nucleation is favored.

$$\Delta\mu_{cr} = 2\Delta\mu_2 = 2(\gamma_{\text{interlayer}} - \gamma_{\text{molecule-substrate}})$$

(2.6)

So at $\Delta\mu \geq \Delta\mu_{cr}$, only 2D crystals are formed. The transition from 3D to 2D growth is one of the most important morphological criterions for high performance pentacene transistors. Equation 2.6 is the most critical equation to analyze and investigate when considering the 2D vs 3D growth at the dielectric interface. For a fixed difference in chemical potential for pentacene deposited on the different surfaces, (i.e., same substrate temperatures and deposition rates) as is the case for the experiments presented in this chapter, Eq. 2.6 shows the growth mode is determined by the strength on the molecule–substrate interaction energy since the interlayer energy is a constant regardless of the underlying OTS. Recall that from the GIXD the lattice of pentacene is identical regardless of the OTS density.

In order to determine the strengths of interactions between pentacene and the various OTS monolayers with varying densities, the experimental conditions (deposition rate and substrate temperature) can be used to determine $\Delta\mu$ and thus by studying the nucleation and growth mode, the interaction energies can be determined. It is also important to note that here it is assumed that the chemical potential difference is between the vapor phase of pentacene and an infinitely large crystal of pentacene (at the substrate temperature). The potential difference, $\Delta\mu$, can be expressed [18]:

$$\Delta\mu = \int_{P_c}^{P_v} \frac{\partial \mu_v}{\partial P} dP - \int_{P_c}^{P_v} \frac{\partial \mu_c}{\partial P} dP = \int_{P_c}^{P_v} (V_v - V_c) dP \qquad (2.7)$$

where $P$ is pressure, $P_v$ is the vapor pressure during deposition and is related to the flux of molecules from the source, and $P_c$ is equilibrium vapor pressure of the crystal at the substrate temperature. In Eq. 2.7 the second equality, the partial derivative of chemical potential in phase $i$ with respect to pressure is equal to the molar volume ($V$) of phase $i$ [19]:

$$\left(\frac{\partial \mu_i}{\partial P}\right)_{T,P} = V_i \qquad (2.8)$$

which leads to the rightmost equality in Eq. 2.7. $V_v$ is the molar volume of the vapor and $V_c$ is the molar volume if the crystal. Since $V_v \ggg V_c$, and because the pressure used during vapor deposition is low enough to assume ideal gas behavior ($P \sim 10^{-6}$ torr) where $V_v = RT/P$, Eq. 2.7 can be rewritten:

$$\Delta\mu \approx RT \ln\left(\frac{P_v}{P_c}\right) \qquad (2.9)$$

The equivalent vapor pressure ($P_v$) can be calculated as a function of the deposition rate ($\theta$) and temperature using [36]:

$$\theta\left(\frac{molecules}{cm^2\ s}\right) = 3.51 \times 10^{22} \frac{P_v}{RT} \qquad (2.10)$$

The equilibrium vapor pressure of a crystal ($P_c$) at substrate temperature $T_{sub}$ is given by [19, 37]:

$$P_c = \exp\left(A - \frac{\Delta H_{sub}}{RT_{sub}}\right) \qquad (2.11)$$

where $\Delta H_{sub}$ is the enthalpy of sublimation and $A$ is a constant related to entropy and have been calculated elsewhere ($\Delta H_{sub} = 37.7$ kcal mol$^{-1}$) [37]. Combining Eqs. (2.9–2.11): the chemical potential driving force can be estimated from the heat of sublimation, and input experimental parameters:

$$\Delta\mu \approx \Delta H_{sub} + RT_{sub}[\ln((2\pi MRT_{sub})\theta) - A] \qquad (2.12)$$

where $R$ is the universal gas constant, $T_{sub}$ is the substrate temperature, and $M$ is the molecular weight of pentacene (278.4 g mol$^{-1}$) [18, 37]. For our pentacene deposition conditions (0.3 Å s$^{-1}$ and a substrate temperature of 60 °C), $\Delta\mu$ is calculated from (2.12) to be 0.07 eV.

Finally, Eq. 2.6 can be analyzed and interaction energies for pentacene and the OTS with differing densities ($\gamma_{mol-substrate}$) can be approximated since now $\Delta\mu$ has been calculated, and $\gamma_{interlayer}$ is known. $\gamma_{interlayer}$ is approximately 0.13 eV, as has been well established from quantum simulations and experiments [19].

$$\Delta\mu_{cr} = 2\left(\gamma_{interlayer} - \gamma_{molecule-substrate}\right) \qquad (2.6)$$

$\gamma_{mol-substrate}$ can then be estimated by determining the growth mode from AFM [19]. For OTS-V, $\gamma_{mol-substrate}$ is ca. $\sim 0.08$–0.9 eV since the growth is highly 3D (also can be approximated by plugging into Eq. 2.4), while for the denser OTS films which engender 2D pentacene growth, $\gamma_{mol-substrate}$ is greater than 0.124 eV; this value is an important numerical heuristic to consider. For a crystalline layer of dense OTS like LB-50, the $\gamma_{mol-substrate}$ is actually greater than $\gamma_{interlayer}$ (as will be shown in detail in the next chapter). Nevertheless, what has been demonstrated is how strong the interaction energy between pentacene (deposited at typical conditions, i.e., rates and substrate temperatures) and the substrate must be in order for 2D growth to be possible. The numerical heuristic is important since one could imagine running simulations to determine the estimated pentacene surface interaction energy for a variety of surfaces to determine which ones give rise to suitable interaction energies needed to drive desirable 2D growth at the dielectric interface.

## 2.6  Conclusions

The most commonly used surface (an alkylsilane modified $SiO_2$) for organic transistors was investigated. The importance of phase and order of the organic dielectric surface modification layer for achieving 2D semiconductor film growth and high charge-carrier mobility in pentacene and $C_{60}$, two of highest performing organic semiconductors has been described. AFM and GIXD provide the first complete picture for the effect of both crystalline order and growth mode of the vital first few semiconductor monolayers on OTFT performance. These results give insight into several new and important issues relevant to engineering high performance devices. Specifically, pentacene's (and many other semiconductors') thin-film growth is highly sensitive to the precise nature of the surface. An increase in density of the methyl terminated surface modification layer results in primarily two-dimensional growth of subsequently vacuum-deposited organic semiconductors. These changes in nucleation and growth give rise to a substantial improvement in the charge-transport characteristics in a number of materials, and suggest that this approach is generally important for the optimization of OTFT (as will be shown in Chap. 4). Finally, since the nucleation and growth mode were determined to be critical for OTFT performance, the chemical potential driving force for heterogenous pentacene crystallization was calculated. The interaction necessary to potentially engender 2D was also calculated. This knowledge may be invaluable and could lead researchers to use simulations to screen potential optimum dielectric materials. In the next chapter, the energetics of nucleation and stability of even thinner films of pentacene (prior to coalescence) are investigated in more detail on crystalline and amorphous OTS.

## 2.7 Experimental

### 2.7.1 Materials

Octadecyltrimethoxysilane (OTMS, 95%, purchased from Gelest Inc.) was purified by distillation and octadecyltrichlorosilane (OTCS, 99%, purchased from Gelest Inc.) was used as received. Device substrates consisted of heavily doped Si wafers with 300 nm of thermally grown silicon oxide having a capacitance per unit area $(Ci)$ of 10 nF cm$^{-2}$. Pentacene was purchased from Sigma-Aldrich and sublimed twice prior to usage. $C_{60}$ was purchased from Alfa Aesar (99.5%) and used as received. For ellipsometry and GIXD experiments, silicon wafers with 2–3 nm of native oxide were used. Prior to OTS treatment the wafers were cleaned with piranha (70:30 $H_2SO_4$:$H_2O_2$) for 60 min and then with ozone plasma (Jetlight UVO-cleaner Model 42–100 V) for 10 min.

### 2.7.2 Fabrication of OTS films

LB Films: A OTMS solution (1 mg ml$^{-1}$ in chloroform) was prepared in a nitrogen glovebox and filtered (0.2 µm pore size). The trough (Nima model 612D) was filled with Millipore water (pH = 3) prepared using concentrated hydrochloric acid (38% HCl). The OTMS films were compressed (20 cm$^2$ min$^{-1}$) with respect to change in trough area to the desired surface pressure and then Blodgett-transferred (1 mm min$^{-1}$) to the Si/SiO$_2$ substrate. The substrates were cleaned sequentially with toluene, isopropanol, acetone, distilled water and isopropanol again and then dried using a nitrogen gun (99.9% pure).

### 2.7.3 Characterization

The grazing angle attenuated total reflectance (GATR) spectrum was obtained using a Nicolet 6700 Fourier Transform Infrared Spectrometer (FTIR) using a germanium crystal.

A Sopra Bois–Columbes ellipsometer with a Physike Instrumente laser (He–Ne, $\lambda = 632.8$ nm, angle of incidence of 70°) and detector were used for OTS thickness measurements. Thickness was calculated from $\Psi$ and $\Delta$ values and measured for five areas on the substrate. The following input refractive indices were used: air, $n_0 = 0$; alkylsilane, $n_1 = 1.450$; native silicon oxide, $n_2 = 1.460$, silicon, $n_3 = 3.873$, $k = -0.016$.

The AFM images of organic semiconductors were collected using a Digital Instruments MMAFM-2 scanning probe microscope. Tapping mode AFM was performed on the samples with a silicon tip with a frequency of 300 kHz.

### 2.7.4 High Resolution AFM

The OTS substrates were washed sequentially with ethanol (99.99% pure from Gold Shield Chemical Co.) and milli-Q water (18.3 MΩ) two times at room temperature (296 K) and allow to air dry before characterization using AFM. The OTS substrates were characterized in decahydronaphthalene solution (99.9%, Sigma-Aldrich).The optimum imaging area to visualize local domain is $100 \times 100$ nm under AFM. Thus, the cursor profile and RMS value was obtained using RHK-based imaging processing software at $100 \times 100$ nm area. For statistical analysis, both domain sizes (FWHM), separation (center-to-center and edge-to-edge) and vertical height were measured quantitatively from over ten cursor profiles per image, using characteristic features, at 300 x 300 nm areas. Repeat experiments yielded similar results.

### 2.7.5 Grazing Incidence X-ray Diffraction

GIXD was performed on the samples at the Stanford Synchrotron Radiation Lightsource (SSRL) on beam line 11–3 with a photon energy of 12.73 keV. A 2D image plate (MAR345) with effective pixel size of 150 μm ($2,300 \times 2,300$ pixels) was used to detect the diffracted X-rays. The detector was 400.15 mm from the sample center. The angle of incidence was fixed at 0.1°. The GIXD data was analyzed using FIT-2D and Peakfit software programs.

The resolution for GIXD was calculated using:

$$\Delta Q_{xy} = \frac{2\pi d \tan(2\theta)}{\lambda D} \qquad (2.13)$$

where:
$\Delta Q_{xy}$ = in-plane resolution ($\text{Å}^{-1}$)
$2\theta$ = the scattering angle (degrees)
$\lambda$ = X-ray wavelength (Å)
$D$ = distance between the sample and the detector (cm)
$d$ = sample length (cm)
Solving: $\Delta Q_{xy} = 0.06 \text{ Å}^{-1}$

### 2.7.6 Electrical Characterization

A Keithley 2,400 semiconductor parameter analyzer was used to test p-channel transistors in an ambient atmosphere, and n-channel transistors in a nitrogen glovebox.

The charge carrier mobility ($\mu$) was calculated by fitting the saturation transfer characteristics using:

$$I_{DS} = \frac{WC}{2L} \mu (V_G - V_T)^2 \qquad (2.14)$$

where $I_{DS}$ is the drain current, $W$ is the channel width, $L$ is the channel length, $C$ is the capacitance of the oxide, $V_G$ is the gate voltage and $V_T$ is the threshold voltage.

## Appendix 2.A: Growth of a Kossel Crystal

Due to its significance on the mobility and conductivity of pentacene thin films, this appendix will introduce general concepts related to crystallization from the vapor phase, and the heterogeneous nucleation of 2D versus 3D crystals. A Kossel crystal is one where all the atoms/molecules are assumed to be cubic in geometry [18]. This is the simplest kind of crystal; more complex geometries often lead to equations which are analytically impossible to solve. Comparing with nucleation of a liquid droplet from a supersaturated vapor, the crystallization of solid crystals is more complex due to the various surfaces with their (often) distinct surface energies [18]. Consider a homogeneous (i.e., not on a substrate or surface) 3D Kossel crystal in equilibrium with the vapor phase (constant temperature and constant volume) then the change in Helmholtz free energy ($dF$) is zero and can be expressed:

$$dF = -P_v dV_v - P_c dV_c + \sum_n \sigma_n dA_n = 0 \tag{2A.1}$$

$$-(P_c - P_v)dV_c + \sum_n \sigma_n dA_n = 0 \tag{2A.2}$$

where $P_v$ is the pressure in the vapor phase, $P_c$ is the vapor pressure of the crystal, $V_v$ and $V_c$ are the vapor and crystal volumes, $\sigma_n$ is the surface energy of surface $n$ with corresponding area $A_n$. Equation 2A.2 is the simplified form of Eq. 2A.1 since at equilibrium the total volume is constant (i.e. ($dV_v = -dV_c$)). The volume of a crystal can also be expressed as a sum of volumes of pyramids with heights $h_n$ and areas $A_n$, as suggested by Wulff in 1901 [18, 38]. $V_c$ and $dV_c$ can then be expressed:

$$V_c = \frac{1}{3} \sum_n h_n A_n \tag{2A.3}$$

$$dV_c = \frac{1}{2} \sum_n A_n dh_n \tag{2A.4}$$

To second order, the very small changes to the total volume $dV_c$ can be accounted for by assuming constant area with infinitesimal changes in pyramid height $dh_n$ (see Markov, Ref. [18] for more details). Reinserting into Eq. 2A.2

$$\sum_n \left[ \sigma_n - \frac{1}{2}(P_c - P_v)h_n \right] dA_n = 0 \tag{2A.5}$$

Since each of the changes in area ($dA_n$) are not related, the first term in the bracket must equal zero

$$P_c - P_v = 2\frac{\sigma_n}{h_n} = \text{constant} \tag{2A.6}$$

This is a restatement of Wulff's rule which states: "at equilibrium, the distances of the crystal faces from a point within a crystal (called Wulff's point which can arbitrarily be chosen as the center of the crystal) are proportional to their corresponding specific surface energies of these faces" [18, 38]. This concept is extremely important in determining whether 2D or 3D crystal growth dominates. Since the chemical potential difference is directly related to the difference in pressures of the two phases by the molar volume of the crystal phase ($V_c$), Eq. 2A.6 can also be written in a more convenient form:

$$\Delta u = u_v - u_v = 2\frac{\sigma_n v_c}{h_n} \tag{2A.7}$$

**Fig. 2.9** The relevant
surface/interfacial energies
used to determine the
equilibrium shape of a crystal

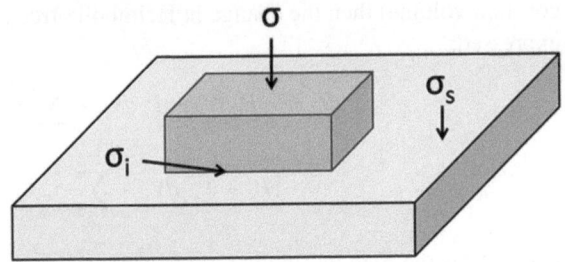

**Fig. 2.9** The relevant
surface/interfacial energies
used to determine the
equilibrium shape of a crystal

This is an important result which mathematically expresses the physical concept that that the supersaturation is the same over the crystal surface, and the growth mode (values of $h_n$) is directly related to the supersaturation. Again, the discussion above was given for a homogenous crystal. For heterogeneous nucleation, which is relevant for organic semiconductor nucleation in OTFTs, Eq. 2A.7 must be slightly modified to include the interaction or adhesion energy ($\sigma_i$) between the crystal and the substrate upon which it is nucleating:

$$\frac{\Delta u}{2v_c} = \frac{\sigma_0 - \sigma_1}{h_n} = \text{constant} \tag{2A.8}$$

where $\sigma_o$ refers to the homogenous case (surface energy); when $\sigma_i$ is zero then the homogenous Eq. 2A.7 is retained. For values where $h_n > 1$, 3D crystals will form, whereas for $h_n = 1$, desirable 2D nucleation prevails. Thus, the term $\sigma_i$, which relates the strength of interaction between the semiconductor and the substrate is a key parameter in determining whether 2D or 3D growth will prevail [18]. The chemical potential driving force and the interfacial energies will determine the growth mode. Define the total change in surface energy upon nucleation on a foreign substrate by $\Delta\sigma$ where:

$$\Delta\sigma = \sigma + \sigma_i - \sigma \tag{2A.9}$$

$\sigma$ is the surface energy of the crystal, $\sigma_i$ is the interfacial surface energy (whose magnitude can be either positive or negative) and $\sigma_s$ is the surface energy of the substrate [18]. There are three basic cases (Fig. 2.9).

*Case 1*: $\Delta\sigma < 0$, this case results when the interaction with the surface is greater than the interlayer interaction energies. Of course in this case, 3D nucleation is prohibited and 2D nucleation can occur at $\Delta\mu = 0$, and even at undersaturation $\Delta\mu < 0$ (provided that $|\Delta\mu| < |A_m\Delta\sigma|$ where $A_m$ is molecular area).

*Case 2*: $\Delta\sigma = 0$ indicates a balanced force between interlayer interaction energy and molecule substrate interactions. This is the general case for nucleation for a material on it crystal of itself (homogenous nucleation). Again in this case, 3D nucleation is thermodynamically impossible, and 2D wetting occurs for supersaturated systems $\Delta\mu > 0$.

*Case 3*: $\Delta\sigma > 0$, or when the system's surface energy increases can give rise to both 2D and 3D growth depending on $\Delta\mu$. This is the general case which was

**Fig. 2.10** A schematic showing how past the critical supersaturation a 3D Kossel crystal must become 2D in order to maintain equilibrium shape from Ref. [18]

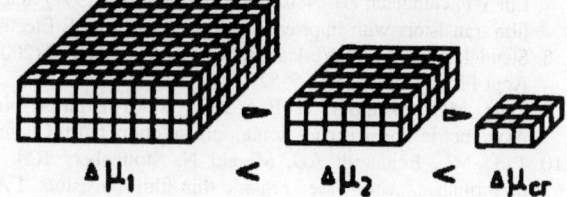

$$\Delta \mu_1 \quad < \quad \Delta \mu_2 \quad < \quad \Delta \mu_{cr}$$

discussed in Chap. 2. The barrier for 3D nucleation, $\Delta G_{3D}^*$, is inversely related to $(\Delta \mu)^2$ (Eq. 2.3) and is possible for all values of $\Delta \mu > 0$. Again 2D nucleation becomes possible only at supersaturations greater than $\Delta \mu_2$, where the change in surface free energy upon nucleation is $\Delta \mu_2 = A_m \Delta \sigma$. This is a logical conclusion, since there must be a driving force greater than the gain in surface energy for nucleation, for the total free energy of the system to decrease. As $\Delta \mu$ increases beyond $\Delta \mu_2$, there exists a critical supersaturation $\Delta \mu_{cr}$ (where $\Delta \mu_{cr} = 2\Delta \mu_2$ at which the $\Delta G_{3D}^* = \Delta G_{2D}^*$) or consequently the height the 3D island is one monolayer high (i.e. a 2D crystal). The extension of Wulff's rule shows that under equilibrium a Kossel crystal will try to maintain its height/length ratio [18, 38] (Fig. 2.10).

In the analysis presented in this chapter on pentacene growth, the chemical potential driving force was fixed, and thus the energetics which determined growth mode are related to the interfacial energies. This allowed for estimation of the interaction energy between pentacene and the different OTS layers. In the following chapter it was determined that on crystalline OTS the pentacene molecule substrate interaction energy is greater than the interlayer interaction energy and in fact this would fall under *case 1* ($\Delta \sigma < 0$) presented above.

The important caveat which must be mentioned is that for systems far from equilibrium (high supersaturations) cannot be addressed using methodology discussed in this chapter, which use thermodynamic models for treating nucleation and crystal shape.

# References

1. Bao Z, Locklin J (2007) Organic field effect transistors. CRC Press Taylor and Francis Group, Boca Raton
2. Dimitrakopoulos CD, Malenfant PRL (2002) Organic thin film transistors for large area electronics. Adv Mater 14:99-+
3. Dinelli F et al (2004) Spatially correlated charge transport in organic thin film transistors. Phys Rev Lett 92:116802
4. Kobayashi S et al (2004) Control of carrier density by self-assembled monolayers in organic field-effect transistors. Nat Mater 3:317–322
5. Heringdorf F, Reuter MC, Tromp RM (2001) Growth dynamics of pentacene thin films. Nature 412:517–520
6. Mayer AC, Ruiz R, Headrick RL, Kazimirov A, Malliaras GG (2004) Early stages of pentacene film growth on silicon oxide. Org Electron 5:257–263

7. Lin YY, Gundlach DJ, Nelson SF, Jackson TN (1997) Stacked pentacene layer organic thin-film transistors with improved characteristics. IEEE Electron Device Lett 18:606–608
8. Steudel S, Janssen D, Verlaak S, Genoe J, Heremans P (2004) Patterned growth of pentacene. Appl Phys Lett 85:5550–5552
9. Tang ML, Okamoto T, Bao ZN (2006) High-performance organic semiconductors: Asymmetric linear acenes containing sulphur. J Am Chem Soc 128:16002–16003
10. Tang ML, Reichardt AD, Miyaki N, Stoltenberg RM, Bao Z (2008) Ambipolar, high performance, acene-based organic thin film transistors. J Am Chem Soc 130:6064
11. Kelley TW et al (2004) Recent progress in organic electronics: materials, devices, and processes. Chem Mater 16:4413–4422
12. Veres J, Ogier S, Lloyd G, de Leeuw D (2004) Gate insulators in organic field-effect transistors. Chem Mater 16:4543–4555
13. Park YD, Lim JA, Lee HS, Cho K (2007) Interface engineering in organic transistors. Mater Today 10:46–54
14. Yang SY, Shin K, Park CE (2005) The effect of gate-dielectric surface energy on pentacene morphology and organic field-effect transistor characteristics. Adv Funct Mater 15:1806–1814
15. Gundlach DJ, Kuo CC, Nelson SF, Jackson TN (1999) Organic thin film transistors with field effect mobility > 2 cm/sup 2//V-s. In: 1999 57th annual device research conference digest (Cat. No.99TH8393). doi:10.1109/DRC.1999.806357
16. Aizenberg J, Black AJ, Whitesides GM (1999) Control of crystal nucleation by patterned self-assembled monolayers. Nature 398:495–498
17. Lee HS et al (2008) Effect of the phase states of self-assembled monolayers on pentacene growth and thin-film transistor characteristics. J Am Chem Soc 130:10556–10564
18. Markov I (2003) Crystal growth for beginners: fundamentals of nucleation, crystal growth and epitaxy, 2nd edn. World Scientific, New Jersey
19. Verlaak S, Steudel S, Heremans P, Janssen D, Deleuze MS (2003) Nucleation of organic semiconductors on inert substrates. Phy Rev B 68:195409
20. Klauk H et al (2002) High-mobility polymer gate dielectric pentacene thin film transistors. J Appl Phys 92:5259–5263
21. Liu SH, Wang WCM, Briseno AL, Mannsfeld SCE, Bao ZN (2009) Controlled deposition of crystalline organic semiconductors for field-effect-transistor applications. Adv Mater 21:1217–1232
22. Lukas S, Sohnchen S, Witte G, Woll C (2004) Epitaxial growth of pentacene films on metal surfaces. Chemphyschem 5:266–270
23. Mannsfeld SCB, Virkar A, Reese C, Toney MF, Bao ZN (2009) Precise structure of pentacene monolayers on amorphous silicon oxide and relation to charge transport. Adv Mater 21:2294
24. Zhang XH, Domercq B, Kippelen B (2007) High-performance and electrically stable C-60 organic field-effect transistors. Appl Phys Lett 91:92114
25. Ulman A (1991) An introduction to ultrathin organic films from Langmuir–Blodgett to self assembly. Academic Press, San Diego
26. Francis R, Louche G, Duran RS (2006) Effect of close packing of octadecyltriethoxysilane molecules on monolayer morphology at the air/water interface. Thin Solid Films 513:347–355
27. Locklin J, Ling MM, Sung A, Roberts ME, Bao ZN (2006) High-performance organic semiconductors based on fluorene-phenylene oligomers with high ionization potentials. Adv Mater 18:2989
28. Porter MD, Bright TB, Allara DL, Chidsey CED (1987) Spontaneously organized molecular assemblies.4. structural characterization of normal-alkyl thiol monolayers on gold by optical ellipsometry, infrared-spectroscopy, and electrochemistry. J Am Chem Soc 109:3559–3568
29. Lee SH, Saito N, Takai O (2007) The importance of precursor molecules symmetry in the formation of self-assembled monolayers. Jpn J Appl Phys Part 1 Regul Pap Br Commun Rev Pap 46:1118–1123

30. Steudel S et al (2004) Influence of the dielectric roughness on the performance of pentacene transistors. Appl Phys Lett 85:4400–4402
31. Chabinyc ML et al (2004) Short channel effects in regioregular poly(thiophene) thin film transistors. J Appl Phys 96:2063–2070
32. Necliudov PV, Shur MS, Gundlach DJ, Jackson TN (2003) Contact resistance extraction in pentacene thin film transistors. Solid-State Electron 47:259–262
33. Dodabalapur A, Torsi L, Katz HE (1995) Organic transistors—2-dimensional transport and improved electrical characteristic. Science 268:270–271
34. Ruiz R et al (2004) Structure of pentacene thin films. Appl Phys Lett 85:4926–4928
35. Yang HC et al (2005) Conducting AFM and 2D GIXD studies on pentacene thin films. J Am Chem Soc 127:11542–11543
36. Ohring M (2001) The material science of thin films, 2nd edn. Academic Press, Orlando
37. Wakayama N, Inokuchi H (1967) Heats of sublimation of polycyclic aromatic hydrocarbons and their molecular packings. Bull Chem Soc Jpn 40:2267
38. Wulff G (1901) On the question of speed of growth and dissolution of crystal surfaces. Zeitschrift Fur Krystallographie Und Mineralogie 34:449–530

# Chapter 3
# The Nucleation, Surface Energetics and Stability of Pentacene Thin Films on Crystalline and Amorphous Octadecylsilane Surface

## 3.1 Introduction

As discussed in the previous chapters, the performance of organic thin film transistors (OTFTs) is strongly dependent on the microstructure of the semiconducting active layer [1–3]. For bottom gated OTFTs, the majority of the current flows within the first few monolayers at the dielectric/semiconductor interface; thus understanding and controlling the growth and nucleation of the semiconductor at this interface is crucial for device optimization [1–3]. The vast majority of the work on the thin film formation of pentacene, one of the highest mobility, and the most widely studied organic semiconductor, was investigated on silicon dioxide (SiO$_2$) [4–9]. While SiO$_2$ is a common (though less so than OTS treated SiO$_2$) dielectric for evaluating organic semiconductors, for future commercial device applications including flexible low cost electronics, the dielectric will be an organic or polymeric material [1, 10–14]. Fundamental understanding and control of pentacene growth on organic surfaces is therefore crucial. As aforementioned, the most common organic surface used for OTFTs is a methyl terminated one wherein the SiO$_2$ is treated with an alkylsilane monolayer like octadecylsilane (OTS) [15–18]. Treating the SiO$_2$ with OTS reduces the surface energy, changes the chemical nature of the surface from polar to non-polar, and from inorganic to organic. OTS modification of SiO$_2$ has also been shown to change organic semiconductor morphology and reduce interfacial hydroxyl groups which are known to be trap states. Most papers reported an increase in OTFT performance (mobility, and on/off) after treatment with OTS [1, 19, 20]. As was discussed in detail in Chap. 2, the molecular order and density of the underlying OTS monolayer was determined to be a critical factor that affects device performance of pentacene OTFTs [18, 21]. Despite the fact OTS treated SiO$_2$ is the most common dielectric surface, there have been few quantitative studies on organic semiconductor growth on OTS modified SiO$_2$.

A. Virkar, *Investigating the Nucleation, Growth, and Energy Levels of Organic Semiconductors for High Performance Plastic Electronics*, Springer Theses, DOI: 10.1007/978-1-4419-9704-3_3, © Springer Science+Business Media, LLC 2012

In Chap. 2 and also later in Chap. 4, it was demonstrated that on a crystalline OTS surface the pentacene OTFT performance was far superior to that on an amorphous OTS layer. The pentacene charge carrier mobility was greater than $3.0 \ cm^2V^{-1}s^{-1}$ on the crystalline OTS and only $0.5 \ cm^2V^{-1}s^{-1}$ on an amorphous OTS [21, 22]. Lee et al. reported a similar phenomenon, with pentacene mobility on ordered OTS typically double that on amorphous OTS [18]. Recall from Chap. 2 that the grazing incidence X-ray diffraction (GIXD) experiments did not indicate any difference in the molecular packing of the pentacene on the differently ordered OTS, but using AFM it was determined that the growth mode and nucleation density were considerably different. A 2D-like pentacene growth and a high nucleation density on the ordered crystalline OTS was observed, whereas on the disordered amorphous OTS, pentacene showed a much lower nucleation density and 3D island growth [22, 23]. It is known that 2D growth at the dielectric is the desirable growth mode for high mobility due less detrimental grain boundaries, but the role of nucleation density is still poorly understood. Counter to common expectations, many of the highest mobility pentacene thin films exhibited a high nucleation density and small crystalline grains at the dielectric interface. [1, 3, 6, 20, 24, 25]. This again indicates that the number of grain boundaries is less important to charge transport than their structure.

In this chapter, the differences in pentacene nucleation density and stability on crystalline versus amorphous OTS will be addressed. It was determined that the role of a dense crystalline methyl terminated surface is primarily to lower the thermo-dynamic barrier of nucleation, which is directly related to the pentacene-surface interaction energy. Moreover, on the crystalline OTS surface, the sub-monolayer pentacene film consisted of high density 2D crystals, which, upon further deposition of pentacene, coalesced into a 2D layer ultimately resulting in a high charge carrier mobility thin film. The work presented in this chapter indicates that nucleation density and growth mode are closely related. While this may seem intuitive, prior to this work, I am not aware of any reports in the field of organic electronics which specifically address the relationship between nucleation density and growth mode. A larger pentacene-surface interaction energy engenders 2D growth and also increases the nucleation density by lowering the thermodynamic barrier to nucle-ation ($\Delta G^*$).

The experimental results presented in this chapter suggest that the differences in pentacene diffusivity on OTS of varying degree of order is not the dominating factor for the differences in nucleation density and growth mode which dictate the charge carrier mobility in completed OTFT. It has been previously suggested that the pentacene diffusivity on OTS of varying density was the major factor gov-erning the nucleation density [18]. Using atomic force microscopy (AFM) and Monte Carlo simulations the relevant energetics of pentacene nucleation and stability on different OTS surfaces were calculated to further demonstrate the importance of the interfacial energy between pentacene and OTS. Sub-monolayer pentacene thin films were 2D and did not reorganize or change with time on a crystalline OTS layer; however on the amorphous OTS surface, the pentacene thin films de-wet from the surfaces over time forming 3D islands. This analysis suggests important considerations in terms of thin film morphology and stability necessary for monolayer OTFTs (see Chap. 6) and sensors [26–28].

## 3.2 Properties of the OTS Monolayers

In order to elucidate the role of methyl surface density on pentacene nucleation and thin film growth, two types of OTS surfaces were prepared. The amorphous, disordered or "liquid-like" OTS was prepared from vapor phase deposition following literature procedures (and Chap. 2) [1, 22, 29]. The crystalline OTS (LB-50) was prepared using the LB technique described in the previous chapter a spin coating technique which is described in detail in the next chapter [21, 22]. Grazing incidence X-ray diffraction (GIXD) was utilized to confirm that the spin-cast OTS was crystalline (which will henceforth be denoted cryOTS) while the vapor phase OTS (henceforth denoted ampOTS) was amorphous. cryOTS packed hexagonally with a 4.2 Å lattice constant, consistent with previous literature reports [30]. The monolayers were further characterized by grazing angle Fourier transform infrared spectroscopy (GATR-FTIR), ellipsometry, and AFM. As expected the area under the absorption curve for the GATR-FTIR spectra was larger for cryOTS compared to ampOTS, again validating that the cryOTS was a more densely packed monolayer. The area under the absorption curve was 1.45 times greater for cryOTS compared to ampOTS, indicating that the cryOTS film was about 1.45 times denser than ampOTS, which corroborates the higher thickness measured by ellipsometry [22]. Also, the characteristic aliphatic stretch modes are shifted to lower wavenumbers for cryOTS, again demonstrating a higher degree of order compared to ampOTS [31]. AFM indicated that the RMS roughnesses of the two films were nearly identical ($\sim 0.2$ nm RMS roughness).

## 3.3 Pentacene Nucleation Density on Different OTS Monolayers

In terms of heterogeneous nucleation the following relevant energetics dominate nucleation and thin film growth: the energetic barrier to diffusion ($E_{diff}$), the energetic barrier to desorption ($E_{des}$), and the energetic barrier required to form a crystal ($\Delta G^*$). The nucleation density ($N_D$) is given by: [21, 32, 33].

$$N_D = R^a \exp\left(\frac{E_i}{kT_S}\right) \tag{3.1}$$

where $R$ is the rate of molecules impinging on the surface, $\alpha$ is a constant related to the critical cluster size, and $E_i$ is the crystal disintegration energy (approximately equal to negative of the crystal formation energy for systems close to equilibrium). Assuming that the relevant energetic barriers to nucleation scale equivalently with the deposition rate (i.e. each has the same exponent), $E_i = (-E_{des} + E_{diff} + \Delta G^*)$, and Eq. 3.1 can be re-written (see Chap. 1). [32, 33]

$$N_D = R^a \exp\left(\frac{-E_{des} + E_{diff} + \Delta G^*}{kT_S}\right) \qquad (3.2)$$

where $k$ is Boltzmann's constant, and $T_s$ is the substrate temperature. Strictly, each of the molecular processes—diffusion, desorption, and nucleation—is also a function of the molecule–substrate interaction energy. The molecule–substrate interaction energy on ampOTS and cryOTS can be approximated by considering the growth mode and was calculated for pentacene in Chap. 2 [23, 25].http://www.rsc. org/delivery/_ArticleLinking/ArticleLinking.cfm?JournalCode=JM&Year=2010& ManuscriptID=b921767c&Iss=Advance_Article - cit19. As aforementioned the pentacene nucleation density is considerably higher on cryOTS than on ampOTS [23]. Figure 3.1 shows AFM images of pentacene deposited on the two different OTS monolayers at different substrate temperatures.

The coverage is lower at higher temperatures indicating significant desorption. The total energetic barrier of nucleation ($E_i$) can be determined by plotting the natural log of nucleation density *versus* inverse substrate temperature (see Fig. 3.2). By fitting the slope in Fig. 3.2, $E_{(cryOTS)}$ was determined to be 0.80 eV and $E_{(ampOTS)}$ was 1.16 eV. There is a smaller total energetic barrier for nucleation on the crystalline OTS.

In the organic transistor literature it is common to assume that the nucleation density scales as $N_D \sim (R/D)$ where $R$ is the deposition rate and $D$ is the diffusivity [18, 34]. In two earlier reports from other research groups on the nucleation of pentacene on ordered versus disordered OTS, this scaling was used to explain why different nucleation densities were seen on different surfaces [18]. From this assumption, for a fixed deposition rate, the nucleation rate is determined *solely* by the inverse of diffusivity. However, this scaling is oversimplified since it does not take into account the energetics of the phase change—i.e. formation of a solid crystal from a supersaturated vapor. The arguments presented prior to the work presented in this chapter took into account $E_{diff}$ and $E_{des}$, but not $\Delta G^*$. These scaling laws, adopted from seminal work in the 1970s and 1980s by Venables and others, do work quite well for a set of inorganic systems on clean surfaces, but do not always translate directly to explain the growth of weakly bound organic thin films where the molecules, and thus their interactions with the surface and each other, are highly anisotropic (see Chaps. 1 and 2) [32–34].

The total energetic barrier to form a crystal is approximated by $-E_i$ ($-E_i = E_{des} - E_{diff} - \Delta G^*$). The difference in desorption and diffusion barriers determines the mean distance a pentacene molecule travels on the OTS [33]. This is true for very early stages of growth. A high barrier for desorption means the molecule is unlikely to desorb because it is strongly bound to the surface. When the barrier for diffusion is low, the pentacene molecules diffusivity is greater. In summary, $E_{des} - E_{diff}$ determines surface mobility [32, 34]. $-\Delta G^*$ defines the

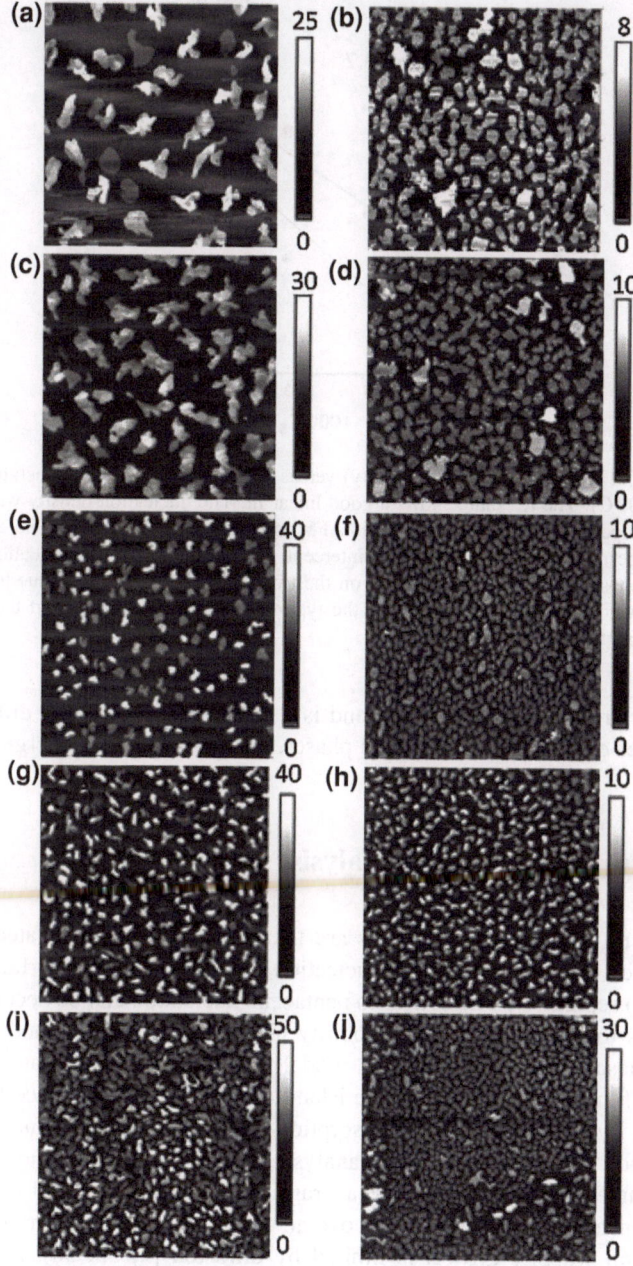

**Fig. 3.1** AFM images of pentacene thin films (nominally 1.5–2 nm measured by a quartz crystal monitor during deposition, deposition rate ∼ 0.2–0.3 Ås$^{-1}$) on cryOTS and ampOTS at different substrate temperatures. **a** 60°C on ampOTS, **b** 60°C on cryOTS, **c** 50°C on ampOTS, **d** 50°C on cryOTS, **e** 40°C on ampOTS, **f** 40°C on cryOTS, **g** 30°C on ampOTS, **h** 30°C on cryOTS, **i** 20°C on ampOTS, **j** 20°C on cryOTS. Each AFM micrograph is 100 μm$^2$

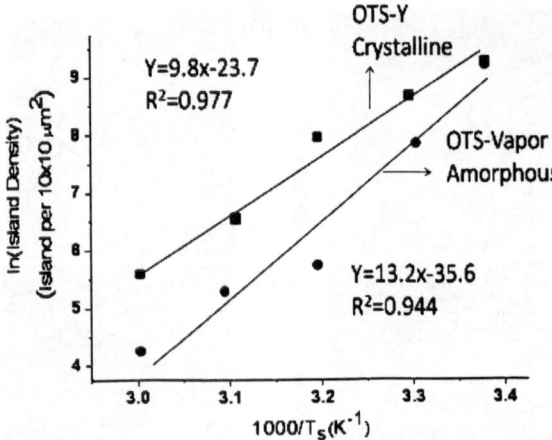

**Fig. 3.2** A plot of the ln (nucleation density) versus $1000/T_s$ for substrate temperatures (20, 30, 40, 50 and 60 °C). The $R^2$ value shows a good linear fit. The nucleation density was calculated using Nanoscope software and analyzing the AFM micrographs. * It should be noted, that since both lines of best fit do not have the same y-intercept, the dependence of the nucleation density on the rate of deposition are slightly different on the two surfaces. This may be due to the critical cluster size which can vary depending on the type of crystal (2D vs. 3D) and the pentacene-substrate interaction energy

energy barrier to crystal formation, and is a strong function of the differences in free energies of the vapor and crystal phases, and the interfacial energies [25, 32].

## 3.4 Capture Zone Model Analysis

To determine which energetic terms are the most important, I treated the energetics related thermodynamics of nucleation and the pentacene surface mobility (diffusivity) separately. The barrier to pentacene nucleation can be decoupled from the barriers to pentacene surface mobility using a capture-zone model for films before coalescence [4, 35, 36]. This model assumes the growth rate of a stable island is a function of the size of the island and the average distance traveled on the surface by a molecule before desorption—i.e. related to $E_{diff}$ and $E_{des}$. Since stable islands are considered in this analysis, the barrier to nucleation ($\Delta G^*$) has been accounted for. Comparing the average spacing of the islands ($L_n$), and the mean distance traveled ($\lambda$), and the overall shape of the crystal grains gives an indication on whether growth is limited by diffusion [4, 35, 36]. $L_n$ and $\lambda$ are shown schematically in Fig. 3.3. $\lambda$ is related to the diffusion, desorption, and the mean residence time on the surface by: [37].

$$\lambda = \sqrt{D\tau} \tag{3.3}$$

**Fig. 3.3** Schematic of
relevant length scales in
capture zone model. The
distance between stable
growing islands is $L_n$, and the
average distance a molecule
diffuses on the surfaces is $\lambda$

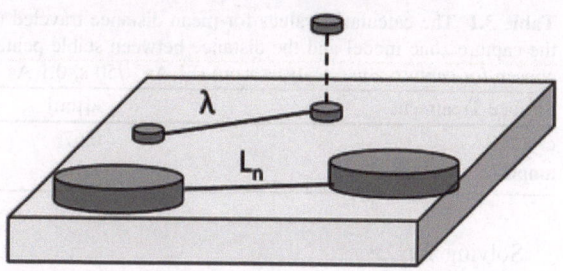

where the diffusivity $D$ is expressed as:

$$D = D_0 \exp\left(\frac{-E_{diff}}{kT_S}\right) \tag{3.4}$$

$D_o$ is the surface diffusion pre-factor, $k$ is Boltzmann's constant, and $T_s$ is the
substrate temperature. The mean residence $\tau$ is related to the surface vibrational
frequency $\upsilon$ and the barrier to desorption by [37] :

$$\tau = \frac{1}{\upsilon}\exp\left(\frac{E_{des}}{kT_S}\right) \tag{3.5}$$

Combining Eqs. 3.4 and 3.5

$$\lambda = \sqrt{\frac{D_0}{\upsilon}\exp\left(\frac{E_{des} - E_{diff}}{kT_S}\right)} \tag{3.6}$$

The surface diffusion pre-factor scales with molecular weight (MW) in terms of
$D_o \sim \text{MW}^{-1/2}$ and surface vibrational frequency scales with substrate temperature as $\upsilon \sim (2\lambda T_s)^{1/2}$. These values are assumed to be identical for pentacene
molecules on both cryOTS and ampOTS surfaces [37]. The distance $\lambda$ is then
directly proportional to $E_{des} - E_{diff}$. Applying capture zone analysis, $\lambda$ can be
approximated by studying the dependence of crystal growth on the deposition rate
and time [4, 36]. The number of molecules in an island as a function of time ($t$) is
given by:

$$\frac{d(n(t))}{dt} = R\lambda P(t) \tag{3.7}$$

where $n(t)$ is the number of pentacene molecules in a crystal, and $P(t)$ is the
perimeter of the islands as a function of time. Using known lattice constants ($a$ and
$b$) for pentacene in the thin film phase (the phase which dominates growth at the
interface) and assuming the islands are circular disks (short cylinders), the area of
the island as a function of time $A(t)$ is:[4, 36]

$$A(t) = \frac{n(t)ab}{2} = \frac{a^2b^2}{8}\left(R^2\lambda^2t^2\right) \tag{3.8}$$

**Table 3.1** The calculated values for mean distance traveled ($\lambda$) by a pentacene molecule using the capture zone model and the distance between stable pentacene islands ($L_n$). The rates/times chosen for capture zone analysis were 0.1 Ås$^{-1}$/50 s, 0.1 Ås$^{-1}$/100 s and 0.1 Ås$^{-1}$/150 s

| Surface Treatment | $\lambda$[μm] | $L_n$[μm] |
|---|---|---|
| cryOTS | 0.67 | 0.66 |
| ampOTS | 1.12 | 1.20 |

Solving for $\lambda$:

$$\lambda = \sqrt{\frac{8A(t)}{a^2 b^2 R^2 t^2}} \tag{3.9}$$

Thus, $\lambda$ can be determined by depositing a material for a given time at a given rate followed by using AFM to measure the average size of the islands. AFM analysis can also be used to determine the mean distance between islands ($L_n$). In each of the three different experiments (for rates and time see Table 3.1), $\lambda_{\text{cry}}$ (for cryOTS) was roughly half that of $\lambda_{\text{amp}}$ (for amorphous OTS). The average values for mean distance between islands ($L_n$) and calculated mean distance traveled are given in Table 3.1.

Though the calculated $\lambda$ is lower for cryOTS compared to ampOTS, this does not necessarily indicate that the diffusivity is lower on cryOTS. When considering the surface mobility, it is imperative to also consider the distance between islands ($L_n$) [4, 36]. On cryOTS, the distance a pentacene molecule travels before joining an existing island is much shorter because the density of stable islands is higher. The distance between stable islands and the average distance travelled are nearly identical on both surfaces. This suggests that the growth rate is not diffusion-limited and, as a result, the diffusivity is probably not dramatically different on cryOTS and ampOTS (i.e. the distance a pentacene molecule "can" travel on cryOTS may very well be larger than on ampOTS but is not experimentally accessible since the nucleation density is also much higher). Reiterating (this is an important point), while differences in pentacene diffusivity may exist on the different surfaces, the effect appears to be negligible compared to other energetics discussed next. The circular shape of the islands is also an indication of non-diffusion limited growth, since diffusion limited growth typically gives rise to highly fractal dendritic islands [6, 20]. Since it appears the surface diffusivity of pentacene on the different surfaces was probably not the major factor governing the nucleation density, the difference should be related to $\Delta G^*$.

## 3.5 Estimating the Barrier to Nucleation

The thermodynamic driving force for nucleation is the difference in chemical potential of the saturated vapor phase ($\mu_v$) and the crystalline phase ($\mu_c$) (see Chaps. 1 and 2), the so-called supersaturation $\Delta\mu$ (where $\Delta\mu = \mu_v - \mu_c$) For a small supersaturations (with respect to thermal energy $kT_s$), the free energy of a 3D crystal with respect to crystal size—number of pentacene molecules $j$—(the special case for a 2D crystal discussed later is given in Chap. 2 and in References 25 and 33) can be described as:[26, 31, 32]

$$\Delta G(j) = -j\Delta\mu + j^{2/3} \sum_i A_i \sigma_i \qquad (3.10)$$

where the first term describes the thermodynamic driving force (the difference in chemical potential of the pentacene vapor phase and crystalline phase) and the second term describes the energetic penalty associated with creating or adding to a new surface with a corresponding surface energy. In the equation above the term $\sigma_i$ corresponds to the surface energy (energy/area) associated with surface $A_i$. For organic crystals formed from pentacene, there are often many facets and different surfaces with different surface energies which make Eq. 3.10 complex. When considering very small islands—where nucleation events occur and the cluster can either disintegrate or grow, Eq. 3.10 becomes intractable. This is due to complexities which arise from both the chemical potential term as well as the surface energy term. First, the chemical potential term becomes hard to define as it is the work required to add a molecule in a phase ($\mu = (\partial G/\partial n)_{T,P}$), and $G(n)$ is not a smooth differential function. As discussed earlier, the Gibbs free energy ($G$) is a macroscopic quantity describing the ensemble energy [32]. Similarly, surface energy is a macroscopic term and thus is ill-defined. Nevertheless, Eq. 3.10 gives the correct macroscopic relationship between free energy, crystal size, and surface energies and is a valid starting point for the analysis of behavior during nucleation. For a fixed rate and substrate temperature, the chemical potential term in Eq. 3.10 can be approximated to be equal for different surfaces (under the assumption that the macroscopic crystal has the same vapor pressure); thus, differences in free energy and growth modes can be attributed to the influences of the surface and the relevant interfacial energies (see Chap. 2) [21, 25].

The barrier to nucleation $\Delta G^*$ can be determined by taking the derivative of $\Delta G$ with respect to the number of molecules $j$, and setting the expression equal to zero:

$$\left( \frac{\partial \Delta G(j)}{\partial j} \right)_{T,P} = 0 \qquad (3.11)$$

At $\Delta G(j) = \Delta G^*$, $j = i$, where $i$ is the so-called critical cluster size [32]. $\Delta G^*$ is the barrier to nucleation where the surface energy effects are maximized. Addition of more molecules to the cluster increases the enthalpic interactions and lowers the total free energy, and thus the intermolecular effects dominate the surface effects, creating a stable island. A first approximation to the barrier to nucleation is obtained by considering the macroscopic surface energies of the various facets in conjunction with molecule surface interaction energies [25]. This approach, adopted by Verlaak and co-workers, can be used to approximate the barrier to nucleation as functions of the chemical potential, the surface energy of each facet, and the molecule–substrate interfacial energy. The barrier to 2D nucleation is given by (see Chap. 2 for barrier to 3D nucleation used to calculated $\Delta G_{3D}^*$ for nucleation on ampOTS) [25]:

$$\Delta G^* = \frac{2(\Psi_{010} + \Psi_{100})(\Psi_{110} + \Psi_{1-10}) - 2\Psi_{101}^2 - 2\Psi_{100}^2 - \Psi_{110}^2 - \Psi_{1-10}^2}{2\Delta\mu - \Psi_{mol\_sub}}$$

(3.12)

where $\psi_i$ (hlk) refers to total energy of surface $i$ (the specific area of surface $i$ times the specific surface energy (see Fig. 2.8). The values for the various pentacene crystal surface energies have been calculated elsewhere experimentally and computationally [9, 25]. The chemical potential driving force for the deposition conditions used in our experiments (vapor pressure $10^{-6}$ Torr, rate 0.3 Ås$^{-1}$, and substrate temperature 60°C) was 0.07 eV. This low driving force further supports using thermodynamic models to treat nucleation. The molecule substrate interaction $\psi_{mol\_sub}$ were also estimated in Chap. 2 and for cryOTS is approximately $\sim 0.14$ eV, and for ampOTS is $\sim 0.09$ eV (Chap. 2). Notice that the molecule substrate interaction energy is very important in determining $\Delta G^*$. The greater the interaction energy, the lower the barrier to nucleation, and the greater the tendency for 2D growth. Substitution of these quantities in Eqs. 2.4 and 3.12 yields: $\Delta G^*_{cryOTS} = 1.85$ eV and $\Delta G^*_{ampOTS} = 2.25$ eV. The difference in barriers to nucleation is thus $\sim 0.40$ eV which is a significant portion of the experimentally determined value of 0.36 eV for the difference in total crystal formation energy. This indicates that the dominant energetic term for pentacene nucleation density on amorphous versus crystalline OTS is in fact the barrier to nucleation—not surface diffusivity. The barrier to nucleation is considerably smaller on cryOTS, which leads to the formation of more pentacene nuclei. Since the density of surface methyl groups on the cryOTS surface and thus the density of OTS molecules itself is $\sim 1.45$ times greater than on ampOTS, the magnitude of $\psi_{mol\_sub}$ is correspondingly larger (see Eq. 3.12). Heterogeneous nucleation can then be described in terms of homogeneous nucleation and a wetting factor related to the overall interactions between the crystallizing molecule and the substrate surface (Eq. 3.13) [37].

$$\Delta G^* = \Delta G_{homo}^* \left\{ \frac{2 - 3\cos\theta + \cos^3\theta}{4} \right\}$$

(3.13)

The left side of the equation is the overall barrier for heterogeneous nucleation and the first term on the right ($\Delta G^*_{homo}$) is the energetic barrier for homogenous nucleation. The second term is a wetting factor which describes the molecular interactions where $\theta$ is the contact or wetting angle which is between 0 and 180° (this is assuming a spherical cap geometry, rigorously pentacene does not form a spherical cap geometry and thus there would be some corrections to the $\theta$ in a modified Young's law) [33]. In the case of complete wetting or $\theta = 0°$, there is no barrier to heterogeneous nucleation. If $\theta = 180°$, the substrate imparts no energetic benefit (i.e., does not catalyze nucleation at all) and the heterogeneous and homogeneous nucleation barriers are equivalent. Figure 3.4 shows AFM images of nominally 3 nm thin films grown on ampOTS and cryOTS. It is clearly evident that on the cryOTS the pentacene film tends to wet the surface and has a higher surface coverage, which again is indicative of a low barrier to heterogeneous nucleation.

On ampOTS 3D nuclei are formed leading to partial surface coverage, and the apparent contact angle is far from 0°, again illustrating a higher barrier to heterogeneous pentacene nucleation compared to cryOTS.

## 3.6 Pentacene Thin Film Stability on Crystalline and Amorphous OTS

The differences in interaction energies also greatly affect the stability of pentacene films on cryOTS and ampOTS. As a crystal of pentacene is formed on a surface, new surfaces with associated interfacial free energies are created between the crystal and the vapor $A_{\text{Pentacenei}} \cdot \gamma_{\text{Pentacenei}}$, where $A_{\text{Pentacenei}}$ is the area of facet $i$ and $\gamma_{\text{Pentacenei}}$ refers to the corresponding interfacial surface energy between facet $i$ and the vapor. There is also an interfacial energy between the crystal and the substrate $A_{\text{Pentacene\_OTS}} * \gamma_{\text{Pentacene\_OTS}}$ (where $A_{\text{Pentacene\_OTS}}$ refers to area of the pentacene in contact with the OTS and $\gamma_{\text{Pentacene\_OTS}}$ refers to the corresponding interfacial energy between pentacene and OTS). Finally there is a loss of interfacial surface energy between the substrate and the vapor ($-A_{\text{Pentacene\_OTS}} \cdot \gamma_{\text{OTS}}$) where $A_{\text{Pentacene\_OTS}}$ refers to the area of the pentacene crystal which is now covering the OTS, and $\gamma_{\text{OTS}}$ is the interfacial surface energy between the OTS surface and the vapor. If the system remains close to thermodynamic equilibrium conditions during the growth, one would expect thin film stability and would not expect any changes in the morphology post deposition (system is at global minimum in terms of free energy).

Recently, Yoshikawa et al. showed that thin films of pentacene (1–3 monolayers) are unstable on amorphous OTS surfaces, while they are very stable on bare SiO$_2$ [38]. They attributed this to the surface energy of SiO$_2$ (61.4 mJ m$^{-2}$) being considerably higher than that of pentacene, whose lowest energy and largest in terms of area is the 001 surface, with a surface energy of 49.7 mJm$^{-2}$. The authors argued that there is no driving force for reorganization—i.e., the system is at equilibrium [39]. In their work, on the lower surface energy of OTS ($\sim 28$ mJm$^{-2}$ for disordered OTS) the pentacene domains aggregated forming 3D islands (within 48–72 h), and by doing so exposed the lower surface energy OTS and minimized the higher surface energy pentacene 001 surface [40]. In this work, a similar phenomenon was observed for thin films of pentacene grown on ampOTS. Over time, the thin films would aggregate and form 3D structures. However, on the cryOTS (surface energy $\sim 24$ mJ m$^{-2}$), despite having a lower

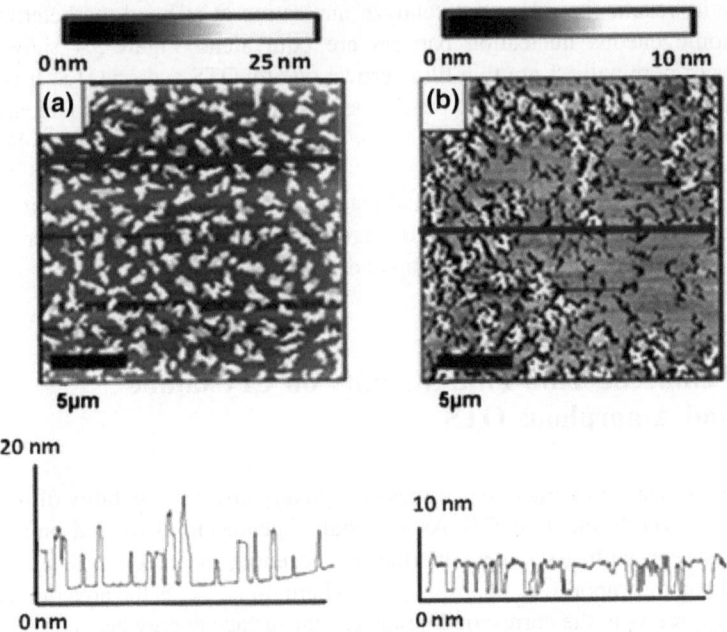

**Fig. 3.4** Atomic force micrographs (AFM) of nominally 3 nm thick pentacene thin films deposited at a rate of. 0.3 $\mathrm{\mathring{A}s^{-1}}$ at a substrate temperature of 60°C on **a** ampOTS and **b** cryOTS. Each AFM is 400 $\mu m^2$ in area. The line profile is provided below each AFM

surface energy than the ampOTS, the pentacene thin films did not undergo reorganization. This vividly illustrates that surface energies are not an adequate metric to discuss the energetics in systems comprised of extended anisotropic particles like organic molecules. In such systems, the magnitude of the interaction energies between the parts of a surface exposed to another phase have very little to do with the energy stored in the surface. Rather, the interaction energy terms of all interfaces need to be considered. In Fig. 3.5 the AFM of thin films of pentacene (nominally 1.5 nm) on ampOTS and cryOTS and a schematic of the relevant surface energies and the reorganization are shown.

The absence of reorganization on cryOTS despite the lower surface energy compared to ampOTS is related to the strong intermolecular interactions between pentacene and the cryOTS surfaces. In order to discuss the stability of a 2D pentacene film, consider the following simplified scenario which is schematically shown in Fig. 3.5e. Assume that a certain amount of pentacene can either be invested in the formation of a large single-layer island (2D case, No.1 in Fig. 3.5e or a two-layer stack that covers half the area on the OTS substrate (3D case, No.2 in Fig. 3.5e). As noted from the AFM the width of a pentacene island is typically much wider ($\sim 0.25$ μm) than the height ($\sim 2$–6 nm). From GIXD investigations it is known that pentacene thin films nucleate with their a–b plane on the substrate surface plane, i.e., {001} facet up. Thus, the facets other than the {001} facets

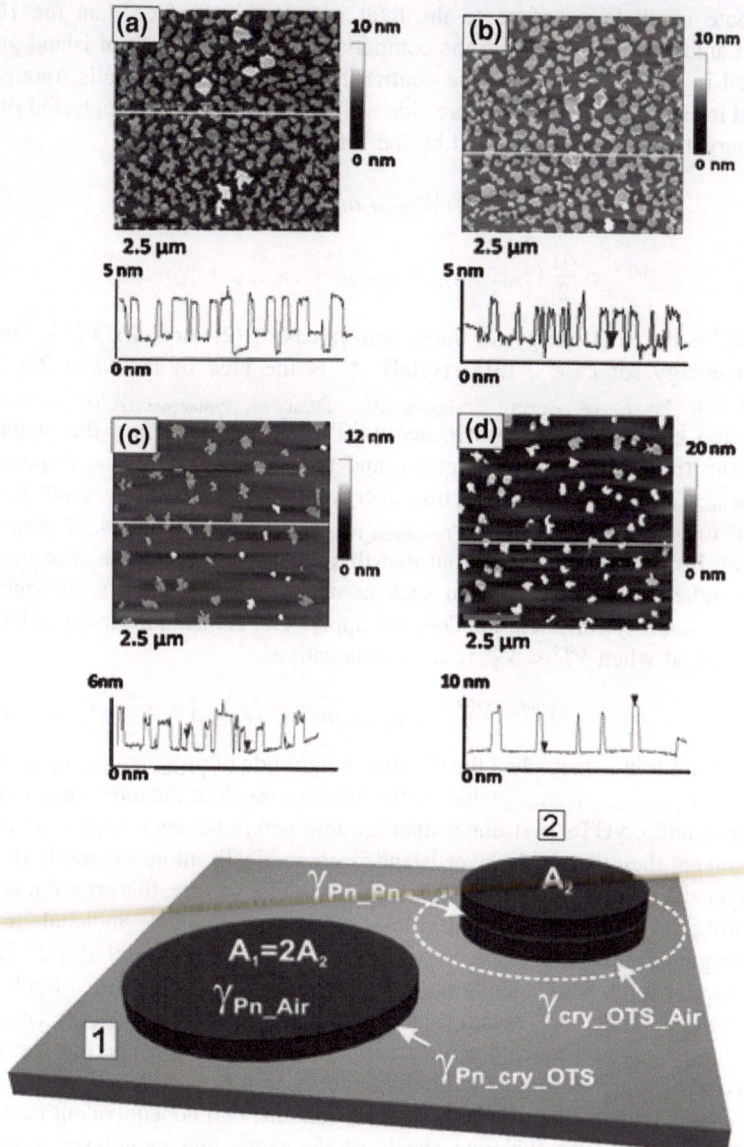

**Fig. 3.5** AFM of thin films of pentacene (nominally 1.5 nm) on cryOTS a) right after deposition, b) after ∼ 48 h and on c) ampOTS right after deposition, d) AFM after ∼ 48 h; e) a schematic showing the relevant interfacial energies in pentacene islands on an OTS substrate. Two possible growth types are discussed: the same amount of pentacene is invested in either (1) a single 2D island or (2) a small 3D island with two stacked layers

contribute much less surface to the total island/grain surface than the {001}-faceted areas. Consequently, in the comparison of the two types of island growth depicted in Fig. 3.5, the energetic contributions from the side walls (though the stacked islands have $\sqrt{2}$ times more side wall area) can be safely neglected [9, 25]. The energetic balance is captured by the following equations:

$$V_1^{tot} = A_1 \left( \gamma_{Pn-cryOTS} + \gamma_{Pn-Air} \right) \tag{3.14}$$

$$V_2^{tot} = \frac{A_1}{2} \left( \gamma_{Pn-cryOTS} + \gamma_{Pn-Pn} + \gamma_{Pn-Air} + \gamma_{OTS-Air} \right) \tag{3.15}$$

here, $V_1^{tot}$ is the total energy for the system in case 1 (2D crystal), $V_2^{tot}$ is the total system energy for case 2 (3D crystal), $A_1$ is the area of the large 2D island (Fig. 3.5e); $\gamma_{Pentacene\_cryOTS}$, $\gamma_{Pentacene\_Air}$, $\gamma_{Pentacene\_Pentacene}$, $\gamma_{OTS\_Air}$ are the interaction energies of the pentacene/cryOTS, the pentacene/air, the pentacene/pentacene (interlayer binding energy), and the OTS/air interfaces, respectively. The $\gamma_{Pentacene\_Air}$, $\gamma_{OTS\_Air}$ interaction energy terms are negligibly small in comparison to $\gamma_{Pentacene\_cryOTS}$ and $\gamma_{Pentacene\_Pentacene}$ ($\sim$ three orders of magnitude smaller). It needs to be pointed out that this is not necessarily the case in atmospheres other than air or vacuum such as organic solvent vapors, in which the following stability consideration does not apply. The 2D island is more stable than the 3D island when $V_1^{tot} < V_2^{tot}$ (i.e. more negative).

$$V_1^{tot} < V_2^{tot} : \left| \gamma_{Pn-cry-OTS} \right| > \gamma_{Pn-Pn}| \tag{3.16}$$

This condition is met when the absolute magnitude of $\gamma_{Pentacene\_cryOTS}$ is greater than that of $\gamma_{Pentacene\_Pentacene}$ which is the intuitive result: if the interaction between pentacene and cryOTS is stronger than the interaction between pentacene in subjacent layers then the single layer island is energetically more favorable than the two-layer stack. Though derived from a much simpler picture, this criterion is again comparing the magnitudes of molecule–substrate and molecule–molecule interaction energies, similar to the one derived by Markov and Verlaak et al. [25, 32].

Molecular force field calculations (non-bonded part of OPLS force-field), were performed in order to obtain approximate values for $\gamma_{Pentacene\_cryOTS}$ and $\gamma_{Pentacene\_Pentacene}$ and to complement/validate the interaction energies estimated from AFM in Chap. 2 [40]. The binding energy of a pentacene monolayer dimer (a dimer was chosen because the 001 plane contains two nonequivalent pentacene molecules) on circular substrate sheets of the pentacene monolayer motif and crystalline OTS was estimated by Monte Carlo sampling over all possible positions within the respective substrate unit cells and azimuthal angles until convergence was obtained. [8] The necessary crystal structure of OTS in the hexagonal 4.2 Å unit cell was obtained from packing calculations with the same force field. The resulting calculated interaction energies are $\gamma_{Pentacene\_cryOTS} = -0.139$ eV and $\gamma_{Pentacene\_Pentacene} = -0.136$ eV. This calculated value for $\gamma_{Pentacene\_cryOTS}$ is in very good agreement with an earlier estimation based on AFM measurements discussed in Chap. 2 [21]. From the GATR-FTIR spectrum, the density of cryOTS/ampOTS is $\sim 29/20$ ($\sim 1.45$). If one assumes that the $\gamma_{Pentacene\_cryOTS}$ value scales with the surface density of the OTS molecules, the value $\gamma_{Pentacene\_ampOTS}$ can be calculated as $(1/1.45)\, \gamma_{Pentacene\_cryOTS} = 0.096$ eV. These values readily explain why pentacene films on crystalline OTS are stable because again:

$$\gamma_{Pn-cryOTS} < \gamma_{Pn\_Pn}\left( |\gamma_{Pn_{cryOTS}}| > | > |\gamma_{Pn_{Pn}}| \right) \tag{3.17}$$

The interfacial energy between the semiconductor and the surface is a critical consideration of OTFTs especially for emerging device architectures where the semiconducting active layer is only a few monolayers [2]. If the interfacial energy is not sufficiently strong, the film will reorganize and the device performance will rapidly decay with time. Current work on very thin OTFTs for monolayer transistors or sensors has relied on depositing the organic semiconductor onto ultra-clean, high surface energy $SiO_2/Si$ substrates [41]. Mono- or bilayer OTFTs are attractive since less material is used lowering device fabrication cost and may be promising candidates for ultra-selective sensors (see Chap. 6). If the OTFT sensor is only a monolayer, extreme sensitivity can be achieved since the analyte molecules will interact directly with the current carrying semiconductor molecules [42]. The high surface energy of $SiO_2$ provides stability against semiconductor reorganization, but unfortunately the same issues discussed earlier (including low performance and rigidity) persist. However, on the highly crystalline OTS monolayer the semiconductor interacts strongly with the terminal methyl group allowing for stable thin film structures which should not undergo reorganization (more on this in Chap. 6).

## 3.7 Conclusions

In this chapter the relevant energetics of pentacene on methyl terminated surfaces were described. Also to complement and expand upon Chap. 2, a more in depth analysis on the nucleation density and stability of pentacene thin films on OTS-modified surface was developed. The crystalline OTS surface increases the pentacene nucleation density by appreciably lowering the thermodynamic barrier to nucleation and not by changing the surface diffusivity. Another fortuitous consequence of the favorable pentacene–cryOTS interaction is improved pentacene thin film stability. It appears that the interfacial energy scales with the SAM density. These findings help to elucidate pentacene thin film formation and growth mode on OTS, the most common surface used for organic transistors, and describe how the dielectric surface should be engineered for higher performance OTFTs, as well as reliable TFTs based on a monolayer of organic semiconductor (Chap. 6). These finding are also potentially useful for sensor applications which will be briefly discussed in subsequent chapters.

# References

1. Bao Z, Locklin J (2007) Organic Field Effect Transistors. CRC Press Taylor and Francis Group, Boca Raton
2. Dinelli F. et al. (2004) Spatially correlated charge transport in organic thin film transistors. Phys Rev Lett 92:116802
3. Dodabalapur A, Torsi L, Katz HE (1995) Organic transistors—2-dimensional transport and improved electrical characteristics. Science 268:270–271
4. Brinkmann M, Pratontep S, Contal C (2006) Correlated and non-correlated growth kinetics of pentacene in the sub-monolayer regime. Surf Sci 600:4712–4716
5. Gundlach DJ, Lin YY, Jackson TN, Nelson SF, Schlom DG (1997) Pentacene organic thin-film transistors—Molecular ordering and mobility. IEEE Electron Dev Lett 18:87–89
6. Heringdorf F, Reuter MC, Tromp RM (2001) Growth dynamics of pentacene thin films. Nature 412:517–520
7. Kelley TW, Muyres DV, Baude PF, Smith TP, Jones TD. (2003) High performance organic thin film transistors. Organic and Polymeric Materials and Devices. Symposium (Mater. Res. Soc. Symposium Proceedings Vol.771), 169–179ⁱxiii + 409
8. Mannsfeld SCB, Virkar A, Reese C, Toney MF, Bao ZN (2009) Precise structure of pentacene monolayers on amorphous silicon oxide and relation to charge transport. Adv Mater 21, 2294–2298
9. Northrup JE, Tiago ML, Louie SG (2002) Surface energetics and growth of pentacene. Phys Rev B (Condensed Matter and Materials Physics) 66, 121404–121407
10. Choudhary D, Clancy P, Bowler DR (2005) Adsorption of pentacene on a silicon surface. Surf Sci 578:20–26
11. Dimitrakopoulos CD, Malenfant PRL (2002) Organic thin film transistors for large area electronics. Adv Mater 14, 99–117
12. Klauk H, Zschieschang U, Pflaum J, Halik M (2007) Ultralow-power organic complementary circuits. Nature 445:745–748
13. Knipp D, Street RA, Volkel A, Ho J (2003) Pentacene thin film transistors on inorganic dielectrics: Morphology, structural properties, and electronic transport. J Appl Phys 93:347–355
14. Knipp D, Street RA, Volkel AR (2003) Morphology and electronic transport of polycrystalline pentacene thin-film transistors. Appl Phys Lett 82:3907–3909
15. Facchetti, A. et al. (2000) Tuning the semiconducting properties of sexithiophene by alpha,omega-substitution—alpha,omega-diperfluorohexylsexithiophene: The first n-type sexithiophene for thin-film transistors. Angew Chem Int Ed 39, 4547–4551
16. Halik M et al (2004) Low-voltage organic transistors with an amorphous molecular gate dielectric. Nature 431:963–966
17. Klauk H et al (2002) High-mobility polymer gate dielectric pentacene thin film transistors. J Appl Phys 92:5259–5263
18. Lee HS et al (2008) Effect of the phase states of self-assembled monolayers on pentacene growth and thin-film transistor characteristics. J Am Chem Soc 130:10556–10564
19. Chua LL et al (2005) General observation of n-type field-effect behaviour in organic semiconductors. Nature 434:194–199
20. Yang HC et al (2005) Conducting AFM and 2D GIXD studies on pentacene thin films. J Am Chem Soc 127:11542–11543
21. Virkar A et al (2009) The role of OTS density on pentacene and C-60 nucleation, thin film growth, and transistor performance. Adv Funct Mater 19:1962–1970
22. Ito Y et al (2009) Crystalline ultrasmooth self-assembled monolayers of alkylsilanes for organic field-effect transistors. J Am Chem Soc 131:9396–9404
23. Virkar A, Ling MM, Locklin J, Bao Z (2008) Oligothiophene based organic semiconductors with cross-linkable benzophenone moieties. Synth Met 158:958–963

24. Park YD, Lim JA, Lee HS, Cho K (2007) Interface engineering in organic transistors. Mater Today 10:46–54
25. Verlaak S, Steudel S, Heremans P, Janssen D, Deleuze MS (2003) Nucleation of organic semiconductors on inert substrates. Phys Rev B 68:195409
26. Locklin J, Bao ZN (2006) Effect of morphology on organic thin film transistor sensors. Anal Bioanal Chem 384:336–342
27. Roberts ME et al (2008) Water-stable organic transistors and their application in chemical and biological sensors. Proc Natl Acad Sci USA 105:12134–12139
28. Roberts ME, Sokolov AN, Bao ZN (2009) Material and device considerations for organic thin-film transistor sensors. J Mater Chem 19:3351–3363
29. Kobayashi S et al (2004) Control of carrier density by self-assembled monolayers in organic field-effect transistors. Nat Mater 3:317–322
30. Lee SH, Saito N, Takai O (2007) The importance of precursor molecules symmetry in the formation of self-assembled monolayers. Jpn J Appl Phys 1 Regul Pap Brief Commun Rev Pap 46, 1118–1123
31. Porter MD, Bright TB, Allara DL, Chidsey CED (1987) Spontaneously organized molecular assemblies.4. structural characterization of normal-alkyl thiol monolayers on gold by optical ellipsometry, infrared-spectroscopy, and electrochemistry. J Am Chem Soc 109:3559–3568
32. Markov I (2003) Crystal growth for beginners: fundamentals of nucleation, crystal growth and epitaxy 2nd edn. World Scientific, New Jersey
33. Venables JA, Spiller GDT, Hanbucken M (1984) Nucleation and growth of thin-films. Rep Prog Phys 47:399–459
34. Venables JA (1973) Rate equation approaches to thin-film nucleation kinetics. Philos Mag 27:697–738
35. Pratontep S, Brinkmann M, Nuesch F, Zuppiroli L (2004) Correlated growth in ultrathin pentacene films on silicon oxide: effect of deposition rate. Phys Rev B 69:165201
36. Pratontep S, Nuesch F, Zuppiroli L, Brinkmann M (2005) Comparison between nucleation of pentacene monolayer islands on polymeric and inorganic substrates. Phys Rev B 72(8):85211
37. Ohring M (2001) The material science of thin films, 2nd edn. Academic Press, Orlando
38. Yoshikawa G. et al. (2007) Spontaneous aggregation of pentacene molecules and its influence on field effect mobility. Appl Phys Lett 90:251906
39. Lim SC et al (2005) Surface-treatment effects on organic thin-film transistors. Synth Met 148:75–79
40. Damm W, Frontera A, TiradoRives J, Jorgensen WL (1997) OPLS all-atom force field for carbohydrates. J Comput Chem 18.1955–1970
41. Chwang AB, Frisbie CD (2000) Field effect transport measurements on single grains of sexithiophene: role of the contacts. J Phys Chem B 104:12202–12209
42. Huang J, Sun J, Katz HE (2008) Monolayer-dimensional 5,5'-Bis(4-hexylphenyl)-2,2'-bithiophene transistors and chemically responsive heterostructures. Adv Mater 20, 2567–2572

# Chapter 4
# Technological Importance of Crystalline Octadecylsilane Monolayers: Crystalline Monolayers Fabricated by Spin-Casting

## 4.1 Introduction

Many linear alkyl derivatives, such as alkylthiols and alkylsilanes, have been used to modify the chemical and/or physical properties of surfaces. These molecules form self-assembled monolayers (SAMs) on the surface by physical adsorption or covalent bonding of the head group to the substrate and are further stabilized by attractive van der Waals packing of the alkyl chains. These SAMs have been widely investigated with particular attention to systems like thiols on gold or silver, silanes on $SiO_2$, phosphonic acids on $Al_2O_3$ and carboxylic acids on metal oxides [1]. Careful control and engineering of the SAM composition and morphology is important for many applications [1–3].

As aforementioned in the past three chapters, octadecylsilanes (OTS), such as octadecyltrimethoxysilane (OTMS) and octadecyltrichlorosilane (OTCS), have been extensively used to modify the $SiO_2$ dielectric surface in organic field-effect transistors (OTFTs) and such modifications result in an increased charge carrier mobility for a variety of vacuum-deposited and solution-cast organic semiconductors [4, 5]. This modification has been used on the dielectric surface of bottom-gated OTFTs since the nucleation and growth behavior of the organic semiconductors is dramatically influenced by composition, roughness and quality of the dielectric surface [6–8]. For high charge carrier transport, two-dimensional (2D) growth is preferred over three-dimensional (3D) or island growth. Thin films that exhibit 3D growth at the dielectric interface tend to form severe grain boundaries during coalescence which act as trap sites that greatly reduce charge carrier mobility. Two-dimensional layer-by-layer growth gives rise to films where islands are better connected, and charge limiting traps are minimized, engendering higher current flow (see Chaps. 2, 3) [9–13].

As discussed in Chap. 2, using the Langmuir–Blodgett (LB) technique, the phase of OTS was systematically varied, and pentacene and $C_{60}$ semiconductor growth and OTFT performance were investigated [5]. Recall that in the LB

A. Virkar, *Investigating the Nucleation, Growth, and Energy Levels of Organic Semiconductors for High Performance Plastic Electronics*, Springer Theses, DOI: 10.1007/978-1-4419-9704-3_4, © Springer Science+Business Media, LLC 2012

technique, amphiphilic OTS molecules were compressed at the air–water interface and under applied lateral pressure a 2D liquid, solid, and even a crystalline monolayer can be formed. The maximum surface pressure that can be achieved was 55 m $Nm^{-1}$ before the OTS film began to collapse and form multilayers. At a surface pressure at or exceeding 50 m $Nm^{-1}$ a dense, crystalline OTS monolayer was formed [5]. On this crystalline OTS monolayer, hereafter referred to as LB-50 (see Chap. 2), the mobility for pentacene and $C_{60}$ thin films was significantly improved over films deposited on amorphous, disordered OTS SAMs. In fact, their mobility increased systematically with the LB OTS density (Chap. 2) [5].

The amorphous OTS SAMs are typically formed using conventional deposition methods, such as solution immersion or vapor deposition (VD). Generally VD gives rise to a smoother surface compared to solution immersion [4, 14]. Silanes, especially trichlorosilanes, polymerize easily in the presence of water and form rough multilayer surfaces. For example, in order to obtain ultra smooth monolayers, Wang et al. applied super dry conditions to fabricate an OTCS monolayer in solution [15]. However, it took 48 h to form a full monolayer, and the solution had to be maintained in ultra-dry conditions. Typically VD is performed under vacuum and the silane is vaporized at elevated temperature (150–200 °C). However, SAMs formed by VD usually have a low density and are disordered [16]. Compared to conventional amorphous OTS SAMs, the crystalline OTS SAM has a higher density of terminal methyl groups which leads to improved "wetting" for a variety of organic semiconductors, i.e., 2D island growth.

However, what was not mentioned in Chap. 2 is that the LB technique is tedious and time-consuming. While it is useful for fundamental research, it is not amenable for real-world applications. The LB technique is not easily scalable to large area processing and films may tear during Blodgett-transfer from the air–water interface creating defect sites [1]. Since a major advantage of organic over inorganic electronics is quick and cheap processing, a new method for producing the crystalline OTS, is highly desirable. In this chapter, a novel spin-coating technique which is quick, simple, and scalable over large areas is described. The thin film properties of the crystalline spin-cast OTS, and the improvement as dielectric modification layers in organic transistors are compared to the vapor deposited films and the LB-50 film is described.

The inspiration for creating highly ordered OTS layers from spin-casting was based on work recently reported by Nie et al. who developed a simple technique to fabricate well ordered monolayers of octadecylphosphonic acid (OPA) on hydrophilic surfaces like aluminum oxide ($Al_2O_3$) [17]. They found that such monolayers can be prepared from solutions of OPA in a non-polar solvent with a dielectric constant between 3 and 5 (trichloroethylene or chloroform) by spin-coating in ambient conditions onto clean oxide surfaces, such as UV/ozone-treated silicon oxide and aluminum-oxide. Phosphonic acid groups can easily form strong covalent bonds with aluminum oxide, but they only physisorb onto silicon oxide. In this chapter, a spin-coating technique which allows for the deposition of a smooth, crystalline OTS SAM with the same quality as the most ordered LB-50 film on silicon oxide is described. The crystalline spin-cast OTS SAM serves as excellent dielectric surface modification layer for OTFTs resulting in very high charge carrier mobilities in a variety of organic semiconductors.

## 4.2 Crystalline OTS Monolayer Deposition and Characterization

Nie et al. used nonpolar solvents with a dielectric constant ($\varepsilon$) around 4, such as trichloroethylene (TCE) or chloroform, to form ordered OPA SAMs on UV/ozone treated hydrophilic oxide surfaces [17]. They hypothesized that ordered OPA monolayers were formed due to the preferential solubility of the hydrophobic tail group in the solvent, and the strong interaction of the polar head group with the hydrophilic substrate surface. They found that the dielectric constant of the solvent is crucial to ensure uniform full-coverage of OPA SAM. Only a small range of solvent polarity results in significant ordering of the polar OPA headgroups on the substrate surface. For the OTS system, solvents with a variety of dielectric constants ($\varepsilon$) were also tested and the corresponding SAM AFM images are shown in Fig. 4.1. The films from hexane ($\varepsilon = 1.89$), toluene ($\varepsilon = 2.38$), and dichloromethane ($\varepsilon = 8.9$) gave rise to polymerized multilayer OTMS films that could not be removed by sonication to give a smooth monolayer (Fig. 4.1). On the other hand, good quality, dense and well ordered monolayers were deposited from chlorobenzene ($\varepsilon = 5.62$) and trichloroethylene (TCE, $\varepsilon = 3.42$). The formation of a crystalline OTS SAM seems to require a similar range of solvent polarity as well ordered OPA SAMs [17].

The proposed formation of an ordered OTS SAM based on the mechanism reported for OPA is illustrated in Fig. 4.2. For OPA, when the solvent dielectric constant is significantly greater than 5, the phosphonic acids start to interact more strongly with the solvent thus disrupting the self-assembly at the substrate-solution interface. However, when the dielectric constant of the solvent is lower than 3, OPA molecules tend to make reverse micelles in solution, which gives rise to incomplete coverage of the OPA SAM. Since alkylsilanes have similar molecular structures, i.e., a non-polar alkyl chain and a polar headgroup, the same mechanism most likely applies. Therefore, based on the above experimental results, TCE was used as solvent for the remaining alkylsilanes described in the rest of this chapter.

In addition to the solvent, the silane solution concentration was also found to be critical for the quality of the SAM film. At low concentrations (<1 mM), incomplete monolayers with isolated islands were formed (Fig. 4.3a). At higher concentrations (>5 mM), multilayers were formed (Fig. 4.3c). Several spin-casting speeds were tested, and 3,000 rpm was found to give the best results. Spin-casting at much lower speeds sometimes led to multilayer formation; faster speeds often led to incomplete coverage. After spin-casting, the substrates were exposed to ammonia or hydrochloric acid vapor for 10 h at room temperature to accelerate

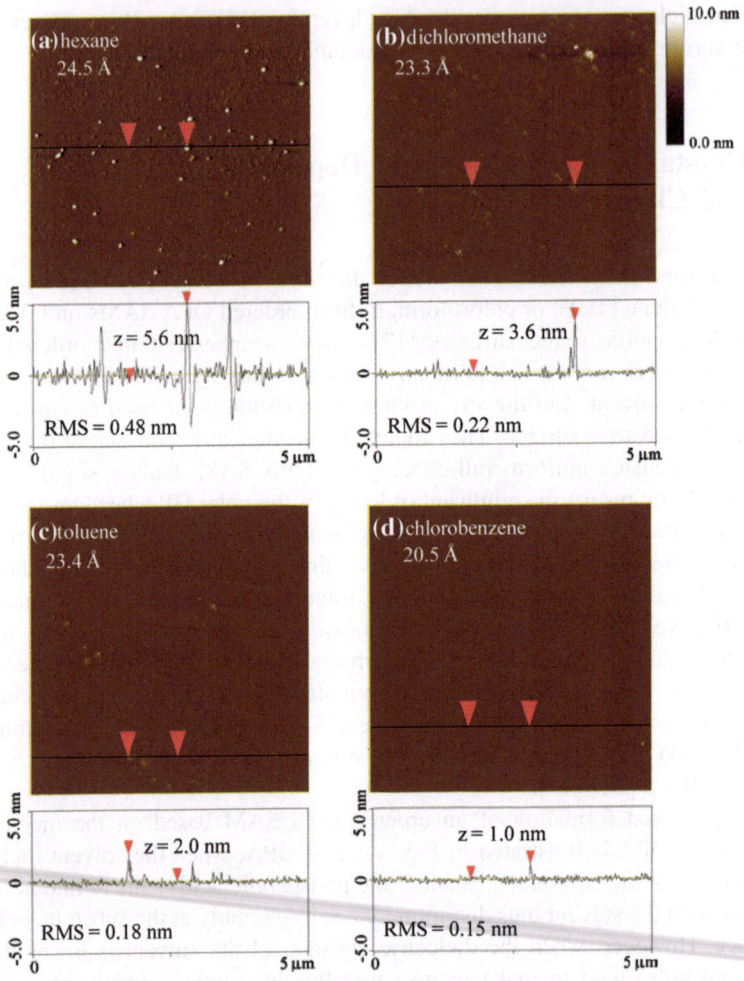

**Fig. 4.1** AFM images and line profiles of OTMS monolayers spin-cast from: **a** hexane, **b** dichloromethane, **c** toluene, and **d** chlorobenzene solutions. The scan area and the height scale of each image are 25 $\mu$m$^2$ and 0–10.0 nm. Z value shows vertical distance between triangles

hydrolysis of alkoxysilanes to and promote bonding to SiO$_2$ surface (see this chapter Appendix for more on the kinetics of SAM formation).

Following the acidic or basic vapor treatment, the substrate was then rinsed or sonicated with solvents to remove any multilayers. If the films did not undergo acid or base treatment, the polymerization and bonding to the surface was often incomplete and the films were easily dissolved or delaminated by organic solvents. If, however, OTCS was chosen as the alkylsilane, acid or base vapor was not necessary to form a crystalline monolayer. The trichlorosilane head group is more reactive and polar than the trimethoxy group. This leads to faster cross-linking.

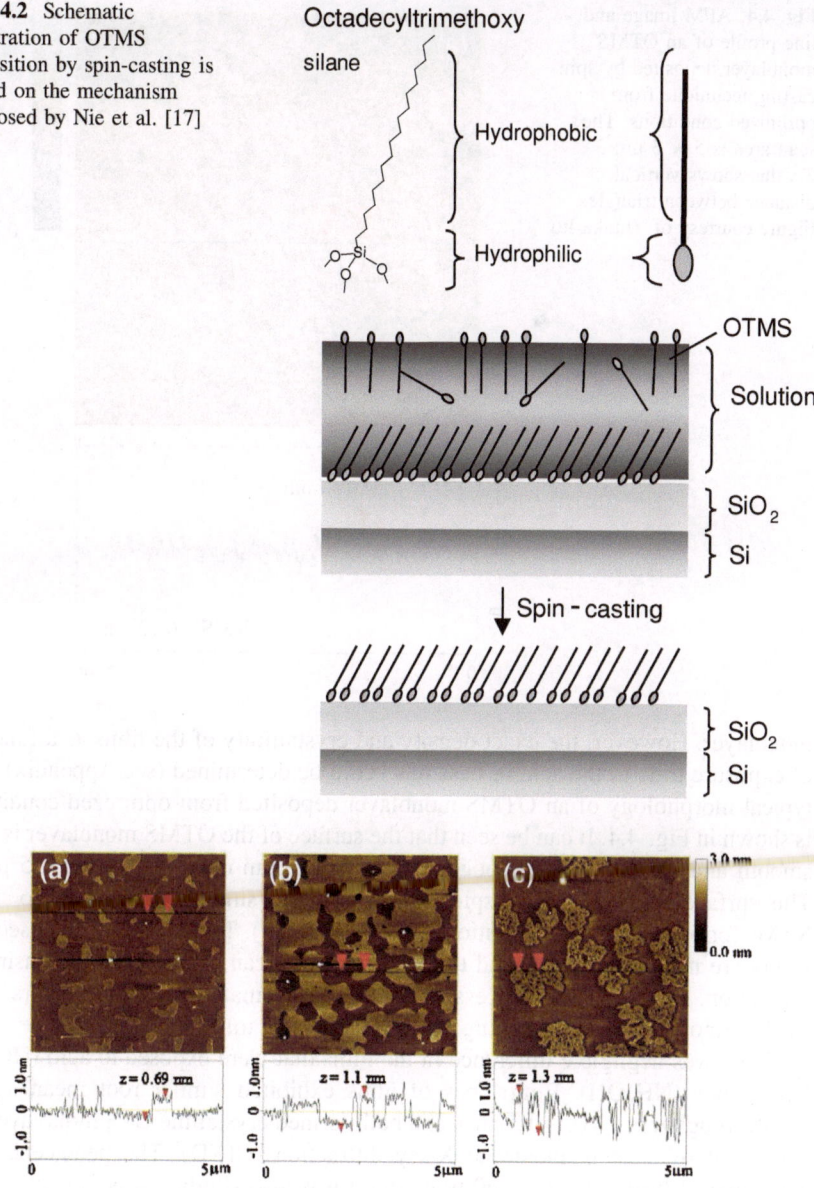

**Fig. 4.2** Schematic illustration of OTMS deposition by spin-casting is based on the mechanism proposed by Nie et al. [17]

Octadecyltrimethoxy silane

Hydrophobic

Hydrophilic

OTMS

Solution

SiO$_2$

Si

Spin - casting

SiO$_2$

Si

(a)  (b)  (c)

3.0 nm

0.0 nm

z = 0.69 nm    z = 1.1 nm    z = 1.3 nm

**Fig. 4.3** AFM images and line profiles of OTMS monolayer films by casting in TCE **a** is <1 mM, **b** is 2–3 mM, and **c** is >3 mM. Image courtesy of Yutaka Ito

Also the driving force for forming the ordered SAM may be higher since the OTCS molecule has a larger difference in polarity between the head group and the alkyl tail. Moreover, kinetic studies later performed show that even 2 h of acid or base can be sufficient to cross-link the OTMS films and form a crystalline

**Fig. 4.4** AFM image and line profile of an OTMS monolayer deposited by spin-casting technique from optimized conditions. The scan area is $5 \times 5 \ \mu m^2$. Z value shows vertical distance between triangles. Figure courtesy of Yutaka Ito

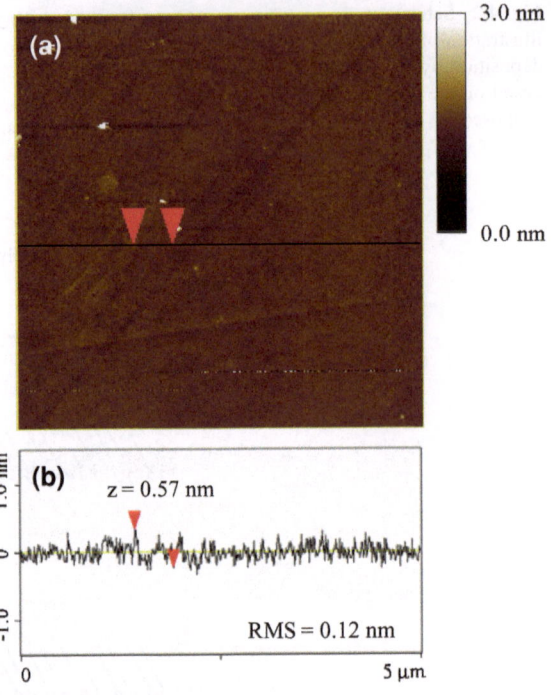

monolayer. However, the exact density and crystallinity of the films as a function of exposure time to the acid or base has yet to be determined (see Appendix). The typical morphology of an OTMS monolayer deposited from optimized conditions is shown in Fig. 4.4. It can be seen that the surface of the OTMS monolayer is very smooth and showed a RMS roughness of 0.1–0.2 nm over large areas (25 $\mu m^2$). This surface produced by the spin-cast technique is smoother compared to silane SAMs formed by vapor or solution deposition [5, 15]. The second and subsequent layers are not covalently bound to the surface and can be removed by rinsing or sonication. In the case of excessive multilayer formation, the multilayers were easily removed by gently wiping the surface with a toluene soaked swab.

There was negligible difference in the films that were exposed to acid (HCl) or base vapor ($NH_4OH$). Both types of films exhibited similar root mean square (RMS) roughness ($\sim 0.1$–0.2 nm), and both formed crystalline OTS monolayers as confirmed by grazing incidence X-ray diffraction (GIXD). The pentacene TFT performance (tested for over 40 transistors) was also similar on both.

Molecules with various alkyl chain length, such as butyltrimethoxysilane (BTMS), octyltrimethoxysilane (Octyl-TMS) and dodecyltriethoxysilane (DTES), were also deposited by the spin-casting technique followed by $NH_4OH$ vapor treatment. It was determined that the difference in polarity between the side chain and the silane group to be vital for formation of well-ordered, dense monolayers: e.g., longer alkyl chain silanes formed better films than shorter alkyl chains. The theoretical molecular lengths, ellipsometric thickness on $SiO_2/Si$ wafer and water

**Table 4.1** Monolayer thickness and static water contact angle

|  | Height (nm) | Contact angle, deg (SD) |
| --- | --- | --- |
| BTMS | 0.7 | 75 |
| Octyl-TMS | 0.6 | 86 |
| DTES | 1.7 | 101 |
| OTSM | 1.8 | 105 |

contact angles of each sample are summarized in Table 4.1. All the molecules except BTMS showed shorter extended lengths than their molecular lengths indicating they may be tilting. However, for OTMS the GIXD data indicate that the molecules are standing up nearly vertically ($\sim 6°$ tilt). The incongruence in the ellipsometric and GIXD data for OTMS is likely due to the difficulty in accurately predicting refractive indices of monolayers on surfaces. The fact the BTMS height was larger the calculated estimated molecular length may be due to experimental inaccuracies with measuring very thin films. BTMS and octyl-TMS SAMs showed low static water contact angles. Much better SAM surface coverage and higher water contact angle were observed for DTES (Table 4.1). As the alkyl chain length increases, the film coverage improves due to increased van der Waals interactions between neighboring molecules [1]. An increase in contact angle without simultaneous increase in roughness is indicative of closer packing of the alkyl chains and a decrease in surface free energy. The contact angles of DTES and OTMS SAMs are similar to those reported previously for smooth alkanes [1, 17].

Additional characterization of spin-cast OTS films were carried out using static water contact angle, and grazing-angle total reflectance Fourier transform infrared spectroscopy (GATR-FTIR). The GATR-FTIR spectrum of the spin-cast OTS film, the LB-50 film, and the amorphous VD OTS film are shown in Fig. 4.5. GATR-FTIR can be used to investigate monolayer density and ordering. The integrated area under the absorption curve is proportional to the monolayer density. The most compressed and highly ordered LB-50 film and spin-cast film showed very similar GATR-FTIR spectrum, which indicate that the spin-cast film has a high degree of order and its mean molecular area is about 20 Å$^2$ molecule$^{-1}$ [5].

In order to determine if the spin-cast monolayer was indeed crystalline, grazing GIXD experiments were performed. GIXD is an ideal tool to investigate the structure of ordered monolayers. In this technique, the thin films are exposed to high intensity synchrotron X-rays under a very shallow angle ($\sim 0.1°$) so that the X-rays penetrate and are scattered not only out-of-plane (perpendicular to the substrate) but also in-plane. Thus for a monolayer, where only in-plane ordering exists, GIXD allows one to determine if in-fact the monolayer is crystalline. If the monolayer is crystalline, it satisfies the Bragg-condition in-plane so that a Bragg rod is observed. Figure 4.8d shows the GIXD images of the spin-cast OTMS monolayer. The diffraction peak of the spin-cast OTMS is unambiguous evidence of the crystalline order of the OTMS monolayer. The diffraction pattern is virtually identical to that of the highest density LB film (50 m Nm$^{-1}$). From the diffractogram the hexagonal lattice constant of the crystalline OTMS was 4.2 Å,

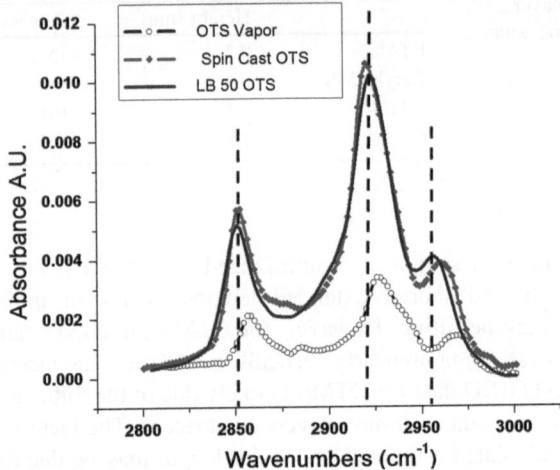

**Fig. 4.5** GATR-FTIR spectrum of the most ordered LB film (with a molecular density of 1 molecule/20 Å² from Refs. [5] and Chap. 2) the spin-cast OTS, and OTS-vapor. The area under the absorption curve can be used to estimate the molecular density. The area under absorption peaks for the LB 50 film and the OTS spin-cast films indicates they are of similar density. The absorbance of the OTS-vapor film is much less which further asserts that it is less dense and less ordered than the crystalline spin-cast and LB OTS films

**Table 4.2** Pentacene TFT performance on spin-cast and vapor deposited OTS

| Surface treatment | $\mu$, cm² V⁻¹ S⁻¹ (SD) | $I_{on}/I_{off}$ (SD) | $V_T$, V (SD) |
|---|---|---|---|
| OTS-V | 0.54 (0.04) | 1.6 (0.02) × 10⁵ | −9.5 (−0.7) |
| Spin cast OTS | 2.9 (0.1) | 4.9 (0.9) × 10⁵ | −13.5 (−3.5) |

which agrees with previous reports for crystalline OTS and the results from Chap. 2 [18]. Formation of dense and crystalline packing by spin casting is unusual for alkylsilanes since common deposition methods usually result in disordered or multilayer films. (See Appendix for more on common errors made using GIXD in organic electronics).

## 4.3 TFT Performance on the Spin-Cast Crystalline OTS Monolayers

Top-contact OTFTs using various organic semiconductors were fabricated and TFT performance was tested to determine the efficacy of the crystalline spin cast OTS SAM as a dielectric surface modification layer. Tables 4.2 and 4.3 shows charge carrier mobility, on/off ratio, and threshold voltage for OTFTs with pentacene (p-channel) or $C_{60}$ (n-channel) active layers on spin-cast OTS and conventional VD OTS [4, 19].

**Table 4.3** $C_{60}$ TFT performance on spin-cast and vapor deposited OTS

| Surface treatment | $\mu$, cm$^2$ V$^{-1}$ S$^{-1}$ (SD) | $I_{on}/I_{off}$ (SD) | $V_T$, V (SD) |
|---|---|---|---|
| OTS-V | 0.27 (0.15) | 7.5 (6.3) $\times$ 10$^5$ | 39.8 (7.5) |
| Spin cast OTS | 4.7 (0.41) | 3.5 (1.2) $\times$ 10$^5$ | 35.6 (6.3) |

Pentacene and $C_{60}$ were chosen for p-type and n-type semiconductors respectively, because they possess exceptionally high field-effect mobilities and are among the most extensively studied organic semiconductors; moreover, direct comparison with the LB-50 film described in Chap. 2 can be made [4, 5]. On spincast OTS SAMs, pentacene and $C_{60}$ OTFTs exhibited mobilities as high as 3.0 and 5.3 cm$^2$ V$^{-1}$ s$^{-1}$, (Fig. 4.6) respectively, while much poorer mobilities of 0.56 and 0.27 cm$^2$ V$^{-1}$ s$^{-1}$ were obtained on vapor-deposited OTS SAMs. The mobilities on the crystalline OTS are among the highest reported for these two organic semiconductors [20, 21].

It is well-known that the charge transport in thin film OTFTs is confined to the first few monolayers at the dielectric-semiconductor interface [22, 23]. To understand the role of the OTS on the charge carrier mobility in more detail, nominally 3 nm of pentacene onto the different OTS-SAM treated substrates. The morphology of these samples was studied by AFM and GIXD. Figure 4.7 shows AFM images of 3 nm pentacene deposited on spin-cast OTMS, highly ordered LB-50 OTS, and vapor OTS. Pentacene grown on vapor (disordered) OTS exhibited undesirable 3D island growth and thus formed a discontinuous film. The AFM line profile for the pentacene grown on VD OTS shows very tall discontinuous 3D islands (Fig. 4.7a). In contrast, the pentacene growth on the crystalline, spin-cast OTS SAMs is very different. The strong interaction between pentacene molecules and the dense methyl terminated substrate resulted in a contiguous 2D sheet with less severe energetic trap states. More in-depth comparison and analysis of semiconductor growth mode on disordered vs. crystalline OTS are given in a Chaps. 2 and 3 [5]. It is important to note that AFM was taken immediately after deposition to avoid film reorganization or degradation of the pentacene deposited on the OTS vapor films (recall there is no pentacene thin film reorganization on the crystalline OTS—see Chap. 3).

The characteristic pentacene (11L), (02L) and (12L) in-plane Bragg rods are seen in the GIXD spectra shown in Fig. 4.8. On the crystalline OTS (Fig. 4.8b and 4.8c), the diffraction from the OTS-SAM can be clearly observed between the (11L) and (02L) pentacene peaks. The lattice constants of pentacene (a = 5.93 Å, b = 7.58 Å, $\gamma \approx 90°$) extracted from the diffraction peaks were nearly identical for both the disordered and crystalline OTS and are similar to those reported for pentacene grown on alkylsilanes (and Chap. 2) [5, 24–26]. The pentacene GIXD spectra (position of peaks in $Q_{xy}$ and $Q_z$) are also similar on all the OTS surfaces. This suggests that the difference in mobility on different OTS surfaces is not due to different pentacene packing motifs. It is also interesting to note that there is an additional diffraction peak at $Q_{xy} = 1.6$ Å$^{-1}$ on vapor OTS

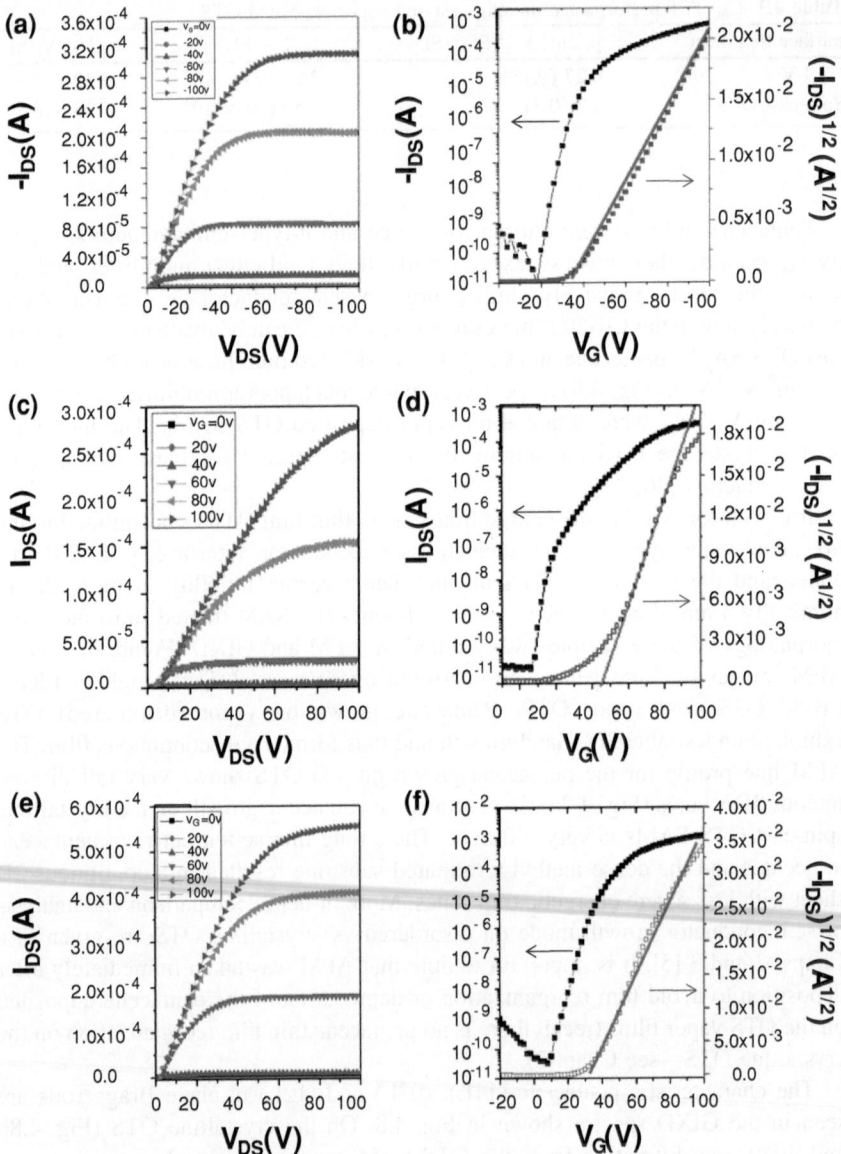

**Fig. 4.6** Representative current–voltage (IV) curves for OTFTs with spin-cast OTS dielectric surface modification layer. **a** Pentacene output IVs, **b** pentacene transfer IVs, **c** PTCDI-C$_4$F$_4$ output IVs, **d** PTCDI-C$_4$F$_4$ transfer IVs, **e** C$_{60}$ output IVs, **f** C$_{60}$ transfer IVs. The red lines in the transfer plots indicate the slope used to calculate mobility. The thickness of the organic semiconductor is about 45 nm measured by a quartz crystal microbalance during deposition

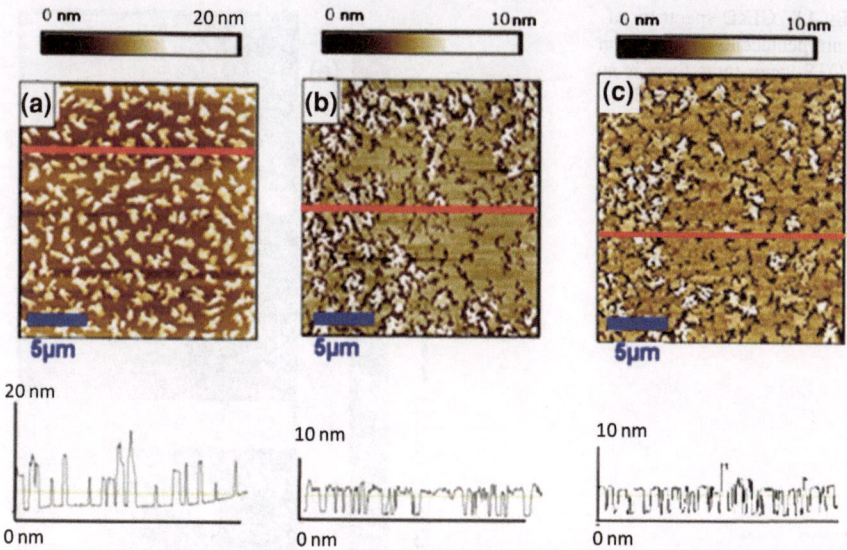

**Fig. 4.7** AFM images of 3 nm, and pentacene deposited on **a** OTS-vapor, **b** spin-cast OTS, and **c** LB 50 OTS

(Fig. 4.8a) which corresponds to a portion of the film exhibiting the bulk pentacene phase (Chap. 2). This partial 3D growth on vapor OTS is consistent with thin film morphology investigated by the AFM (Fig. 4.7). In principle, the full-width at half max (FWHM) of the diffraction peaks can be used to gauge the crystalline quality of the pentacene on various OTS surfaces, but for all the films studied, the FWHM was resolution limited (due to sample size effects see Chap. 5 Appendix)

For the typical n-channel material, $C_{60}$, the densely packed SAM served as an excellent dielectric modification layer and a field-effect electron mobility as high as 5.3 cm$^2$ V$^{-1}$ s$^{-1}$ was achieved. In addition to the morphological effects on semiconductor growth, the densely packed SAM effectively passivates electron traps on SiO$_2$ which can also contribute to the high mobilities especially for electron transporting materials [27–29]. Another other n-channel organic semi-conductors based on perylene tetracarboxylic diimide (PTCDI), which is one of the most promising n-channel candidates due to the high electron affinity and the large π-orbital overlap in the solid state, was also tested. The field-effect mobility of $N,N'$-bis(heptafluorobutyl)-3,4:9,10-perylene tetracarboxylic diimide (PTCDI-C4F7) increased to 1.4 cm$^2$ V$^{-1}$ s$^{-1}$ on the spin-cast OTS compared to 0.72 cm$^2$ V$^{-1}$ s$^{-1}$ for the OTFT device prepared on vapor-treated OTS [19]. A PTCDI compound with two fluorine atoms at the core-aromatic ring also exhibited enhanced mobilities on spin-cast OTMS SAM layer (0.66 cm$^2$ V$^{-1}$ s$^{-1}$) compared to the vapor-treated OTS (0.35 cm$^2$ V$^{-1}$ s$^{-1}$) [3, 4]. The spin cast OTS allowed for

**Fig. 4.8** GIXD spectrum of 3 nm pentacene deposited on **a** OTS-vapor (note there is no diffraction peak observed from the amorphous OTS), **b** spin-cast OTS, **c** LB-50 OTS, **d** spin-cast OTS (with no pentacene) showing that, indeed, the monolayer is crystalline

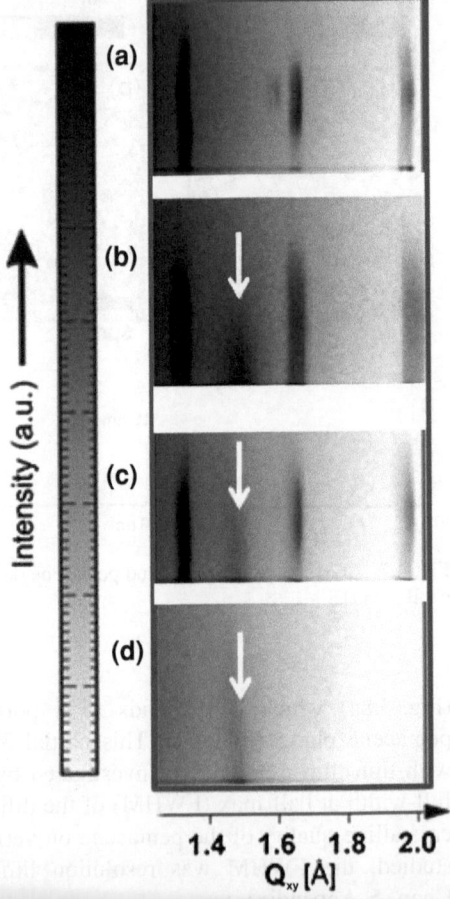

the observation of ambipolar transport for a variety of pentacene derivatives while they could not be observed on vapor OTS (see Fig. 4.9) [27, 30]. These results indicate that the spin-cast OTS SAM layer is useful for the preparation of high performance n-channel and ambipolar OTFTs [14, 31].

The corresponding list of semiconductors whose mobility is presented in Fig. 4.9 is provided below. Electron mobilities are presented in bold, and the threshold voltages are italicized.

| Molecule | $T_{mob}$ (°C) | $\mu$ (cm²/Vs) Mobility on OTSY $V_T(V)$ | $\mu$ (cm²/Vs) OTS–V $V_T(V)$ |
|---|---|---|---|
| 1 | 125 | 1.42 ± 0.03 (+46) | 0.67 ± 0.05 (+27) |
| 2 | 70 | 1.181 ± 0.24 (-5) | 0.67 ± 0.06 (-15) |
| 3 | 125 | 0.86 ± 0.05 (+28) | 0.24 ± 0.03 (+22) |
| 4 | 125 | 0.76 ± 0.06 (+9) | 0.029 ± 0.014 (+6) |
| 5 | 125 | 0.64 ± 0.02 (+14) | 0.31 ± 0.04 (+14) |

(continued)

(continued)

| Molecule | $T_{mob}$ (°C) | $\mu$ (cm²/Vs) Mobility on OTSY $V_T(V)$ | $\mu$ (cm²/Vs) OTS–V $V_T(V)$ |
|---|---|---|---|
| 6 | 70 | 0.595 ± 0.18 (+60) | 0 |
| 7 | 125 | 0.59 ± 0.03 (+28) | 0.31 ± 0.02 (+6) |
| 8 | 60 | 0.391 ± 0.07 (+20) | 0.271 ± 0.04 (+10) |
| 9 | 70 | 0.28 ± 0.03 (+30) | 0.0004 ± $10^{-5}$ (+10) |

(continued)

(continued)

| Molecule | $T_{mob}$ (°C) | $\mu$ (cm²/Vs) Mobility on OTSY $V_T(V)$ | $\mu$ (cm²/Vs) OTS–V $V_T(V)$ |
|---|---|---|---|
| 10 | 125 | 0.28 ± 0.07 (+29) | 0.023 ± 0.010 (+23) |
| 11 | 60 | 0.255 ± 0.03 (-20) | 0.166 ± 0.007 (-5) |
| 12 | 70 | 0.123 ± 0.02 (+40) | 0 |

(continued)

(continued)

| Molecule | $T_{mob}$ (°C) | $\mu$ (cm²/Vs) Mobility on OTSY $V_T(V)$ | $\mu$ (cm²/Vs) OTS–V $V_T(V)$ |
|---|---|---|---|
| | | | 0.00209 ± 0.0006 (+10) |
| 13 | 70 | 0.114 ± 0.02 (+30) | |
| 14 | 125 | 0.038 ± 0.011 (+32) | 0 |

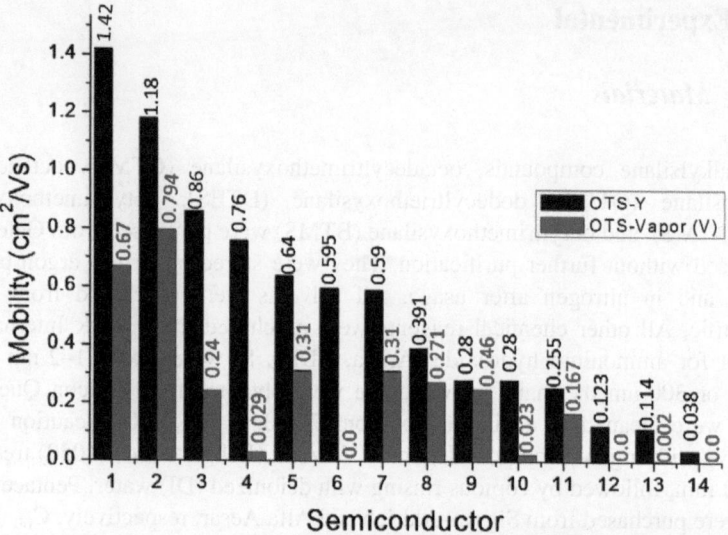

**Fig. 4.9** The average mobility (for at least six devices on each OTS treatment) on crystalline OTS-Y (*Left*) and amorphous OTS-V (*Right*). Both the electron and hole mobility on a variety of semiconductors is higher on the crystalline OTS. See below for the corresponding list of semiconductors (most of these molecules were synthesized by Dr. Ming Lee Tang)

## 4.4 Conclusions

In this chapter, a simple, ambient condition, solution-deposition technique to create crystalline layers of OTS on $SiO_2$ surfaces was described. Modifying the $SiO_2$ dielectric using a crystalline OTS layer compared to the conventional amorphous OTS resulted in much higher charge carrier mobilities for both p- and n-channel vapor deposited organic semiconductors. The improvement in performance was again attributed to the ability to control the semiconductor growth to be the more desirable 2D growth on crystalline OTS leading to well connected highly conductive films, as opposed to 3D growth which is commonly observed on amorphous OTS. The high density and close packing of the terminal methyl groups in crystalline OTS monolayer interact favorably with the semiconductor layer initially deposited and template 2D growth (see Chaps. 2 and 3). Moreover, compared to other techniques for crystalline OTS deposition, such as the Langmuir–Blodgett method, this technique is more amenable to large area processing. Although a 5 inch wafer was the largest we demonstrated, this technique should be scalable to larger areas. The formation of a crystalline OTS layer, which greatly increases performance, from a simple solution deposition process represents an important development for organic electronics.

## 4.5 Experimental

### 4.5.1 Materials

The alkylsilane compounds, octadecyltrimethoxysilane (OTMS), octadecyltri-chlorosilane (OTCS), dodecyltriethoxysilane (DTES), octyltrimethoxysilane (Octyl-TMS), and butyltrimethoxysilane (BTMS) were purchased from Gelest Inc. and used without further purification. They were stored under dry argon prior to usage and in nitrogen after usage. All solvents were purchased from Fisher Scientific. All other chemical reagents were purchased from VWR International except for ammonium hydroxide (Arista, BDH). Si wafers with 1–2 nm native oxide or 300 nm thermally grown oxide were obtained from Silicon Quest inc. They were cleaned in a piranha solution (70:30, $H_2SO_4:H_2O_2$—caution highly reactive with organic compounds) and UV/ozone (Jetlight Model 4050) treatment for 10 min, followed by copious rinsing with deionized (DI) water. Pentacene and $C_{60}$ were purchased from Sigma–Aldrich and Alfa Aesar, respectively. $C_{60}$ (99.5% pure) was used as received, and pentacene was purified twice by zone sublimation before usage. Professor Frank Würthner from the Universität Würzburg provided the PTCDI-$C_4F_4N,N'$-bis(heptafluorobutyl)-3,4:9,10-perylene tetracarboxylic dii-mide (PTCDI-C4F7) and was sublimed prior to usage. Other semiconductors were synthesized by Dr. Ming Lee Tang.

### 4.5.2 Characterization

Static contact angles were measured with an Edmund Scientific goniometer and the probe fluid was milli-Q water. Ellipsometric measurements of SAMs on Si wafers with a native oxide were performed with a Sopra Bois-Columbes ellips-ometer. The light source was a Physike Instrumente He–Ne laser with $\lambda = 632.8$ and the angle of incidence was 70°. The thickness of the SAM was calculated from the measured $\Psi$ and $\Delta$ values using special integrated software (Optrel GbR) with the following parameters: air, refractive index ($n_0$) = 1.0; alkylsilane, $n_1 = 1.450$; native silicon oxide, $n_2 = 1.460$, thickness (d)= 1.77 nm; silicon, $n_3 = 3.873$, $k = -0.016$. The SAM films were assumed to be isotropic and homogeneous.

The atomic force microscope (AFM) images of organic semiconductor thin films and the SAM-treated $SiO_2$/Si substrates were collected using a Digital Instruments MMAFM-2 scanning probe microscope. Tapping mode AFM was performed on the samples with a silicon tip with a frequency of 300 kHz.

The grazing angle attenuated total reflectance (GATR) spectrum was obtained using a Nicolet 6700 Fourier Transform Infrared Spectrometer (FTIR) using a germanium crystal.

Grazing incidence X-ray diffraction (GIXD) experiments were performed at the Stanford Synchrotron Radiation Lightsource (SSRL) on beam line 11–3 with a photon energy of 12.73 keV. A 2D image plate (MAR345) with effective pixel size of 150 μm (2,300 × 2,300 pixels) was used to detect the diffracted X-rays. The detector was 400.15 mm from the sample center. The angle of incidence was kept fixed at 0.1° to maximize the diffracted signal and minimize the background from the substrate scattering. The GIXD data was analyzed using FIT-2D and Peakfit software programs.

### 4.5.3 TFT Device Fabrication

Heavily n-doped silicon substrates with a thermally grown 300 nm silicon dioxide dielectric layer with a capacitance per unit area ($Ci$) of 10 $nF/cm^2$ were used as transistor substrates. For top-contact geometry, the organic semiconductors were deposited at a rate of 0.3–0.6 Å $s^{-1}$ under a pressure of 5.0 × $10^{-7}$ Torr and a substrate temperature of 60 °C for pentacene or 110 °C for $C_{60}$ to a final thickness of 45 nm determined by a quartz crystal monitor in the evaporation chamber. Then, gold electrodes ($\sim 40$ nm in thickness) were deposited using shadow masks with a $W/L$ of 20($W$ = channel width, $L$ = channel length), where $L = 40$–200 μm. The electrical characteristics were obtained at room temperature using a Keithley 4200 (Hewlett-Packard) semiconductor parameter analyzer in air for pentacene or under nitrogen for $C_{60}$. Transfer IV characteristics were obtained with a fixed source-drain voltage of $-100$ V for pentacene TFTs and 100 V for $C_{60}$ TFTs.

## 4.6 General Method for Fabrication of Crystalline OTS SAM from Spin-Casting

General alkylsilane deposition conditions were developed by modifying the procedure reported for octadecylphosphonic acid (OPA) monolayers on hydrophilic substrates by spin-casting [17]. For alkyl-trialkoxysilanes, the solution was dispensed onto $SiO_2$/Si wafers and allowed to partially self-assemble for 10 s prior to spinning coating at 3,000 rpm for 10 s. The substrate was subsequently vapor annealed in ammonia or hydrochloric acid. The best conditions for crystalline OTS is the following: 3 mM OTMS solution in trichloroethylene (TCE) was cast onto a UV/ozone cleaned $SiO_2$/Si wafer to cover the entire surface and was allowed to partially assemble for 10 s; the substrate was then spun at 3,000 rpm for 10 s. Following spin-casting the substrate was put in a closed container with a small vial which contained a few millimeters of ammonium hydroxide solution (28–30% in water) for 10 h at room temperature. The substrates were then rinsed with DI water and sonicated in toluene.

**Fig. 4.10** GIXD diffractogram of OTCS spin-cast monolayer (without any exposure to acid or base vapor). The large arrow points to the diffraction rod due to the underlying OTS being crystalline

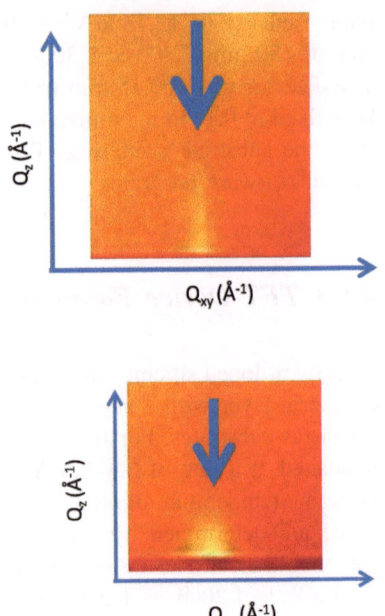

**Fig. 4.11** GIXD diffractogram of OTMS spin-cast monolayer after exposure to HCl for 2 h. The large arrow points to the diffraction rod due to the underlying OTS being crystalline

# Appendix: Kinetics of Self Assembled Monolayer Formation

As mentioned in this chapter, the kinetics of the SAM formation are highly sensitive to the ambient moisture. Initially, we noticed that without acid or base hydrolysis the SAM would not form covalent bonding and could be easily removed by organic solvents. It was determined that for the OTMS (the trimethoxy) version, crystalline SAMs formed after 10 h of exposure to the acid or base vapor. However, initially rigorous kinetic studies were not performed. In subsequent months after discovering the spin-cast OTS technique, and the general importance of crystalline OTS, we sought to study more about the kinetics of SAM formation.

OTMS (octadecyltrimethoxysilane) and OTCS (octadecyltrichlorosilane) monolayers were deposited using the spin-cast technique described in this chapter. The films were then placed under HCl or NH₄OH vapor for either: 0, 2, 4, 6, 8, or 10 h. After the exposure to HCl the films were removed and characterized using grazing incidence X-ray diffraction. Interestingly, the OTCS films did not require any acid or base vapor in order to form a crystalline layer. See Fig. 4.10. The high reactivity of the trichlorosilane group allowed for formation of the crystalline SAM without additional catalysis provided by the acid or base vapor. For OTMS the crystalline SAM could be formed after 2 h of exposure to HCl. There is ongoing work trying to understand more about the formation of crystalline OTS monolayers. The proposed plan includes trying to study in situ the formation of crystalline layers using GIXD (Fig. 4.11).

# References

1. Ulman A (1991) An introduction to Ultrathin organic films from Langmuir–Blodgett to self assembly, 1st edn. Academic Press, San Diego
2. Aizenberg J, Black AJ, Whitesides GM (1999) Control of crystal nucleation by patterned self-assembled monolayers. Nature 398:495–498
3. Porter MD, Bright TB, Allara DL, Chidsey CED (1987) Spontaneously organized molecular assemblies 0.4. Structural characterization of normal-alkyl thiol monolayers on gold by optical ellipsometry, infrared-spectroscopy, and electrochemistry. J Am Chem Soc 109:3559–3568
4. Bao Z, Locklin J (2007) Organic field effect transistors. CRC Press Taylor and Francis Group, Boca Raton
5. Virkar A et al (2009) The role of OTS density on pentacene and C-60 nucleation, thin film growth, and transistor performance. Adv Funct Mater 19:1962–1970
6. Tang ML, Oh JH, Reichardt AD, Bao ZN (2009) Chlorination: a general route toward electron transport in organic semiconductors. J Am Chem Soc 131:3733–3740
7. Tang ML, Reichardt AD, Miyaki N, Stoltenberg RM, Bao Z (2008) Ambipolar, high performance, acene-based organic thin film transistors. J Am Chem Soc 130:6064
8. Veres J, Ogier S, Lloyd G, de Leeuw D (2004) Gate insulators in organic field-effect transistors. Chem Mater 16:4543–4555
9. Ruiz R et al (2004) Pentacene thin film growth. Chem Mater 16:4497–4508
10. Tang ML et al (2006) Structure property relationships: asymmetric oligofluorene-thiophene molecules for organic TFTs. Chem Mater 18:6250–6257
11. Verlaak S, Steudel S, Heremans P, Janssen D, Deleuze MS (2003) Nucleation of organic semiconductors on inert substrates. Phy Rev B 68:195409
12. Virkar A, Ling MM, Locklin J, Bao Z (2008) Oligothiophene based organic semiconductors with cross-linkable benzophenone moieties. Synthet Met 158:958–963
13. Yang HC et al (2005) Conducting AFM and 2D GIXD studies on pentacene thin films. J Am Chem Soc 127:11542–11543
14. Tang ML, Okamoto T, Bao ZN (2006) High-performance organic semiconductors: asymmetric linear acenes containing sulphur. J Am Chem Soc 128:16002–16003
15. Wang YL, Lieberman M (2003) Growth of ultrasmooth octadecyltrichlorosilane self-assembled monolayers on SiO2. Langmuir 19:1159–1167
16. Peanasky J, Schneider HM, Granick S, Kessel CR (1995) Self-assembled monolayers on mica for experiments utilizing the surface forces apparatus. Langmuir 11:953–962
17. Nie HY, Walzak MJ, McIntyre NS (2006) Delivering octadecylphosphonic acid self-assembled monolayers on a Si wafer and other oxide surfaces. J Phys Chem B 110:21101–21108
18. Lee SH, Saito N, Takai O (2007) The importance of precursor molecules symmetry in the formation of self-assembled monolayers. Jpn J Appl Phys Part 1 Regul Pap Br Commun Rev Pap 46:1118–1123
19. Oh JH, Liu S, Bao Z, Schmidt R, Wurthner F (2007) Air-stable n-channel organic thin-film transistors with high field-effect mobility based on N,N '-bis(heptafluorobutyl)3,4 : 9,10-perylene diimide. Appl Phys Lett 91:212107
20. Gundlach DJ, Kuo CC, Nelson SF, Jackson TN (1999) Organic thin film transistors with field effect mobility >2 cm/sup 2//V-s. In: 1999 57th annual device research conference digest (Cat. No.99TH8393). doi:10.1109/DRC.1999.806357

21. Zhang XH, Domercq B, Kippelen B (2007) High-performance and electrically stable C-60 organic field-effect transistors. Appl Phys Lett 91:92114
22. Dinelli F et al (2004) Spatially correlated charge transport in organic thin film transistors. Phys Rev Lett 92:116802
23. Dodabalapur A, Torsi L, Katz HE (1995) Organic transistors—two-dimensional transport and improved electrical characteristics. Science 268:270–271
24. Fritz SE, Martin SM, Frisbie CD, Ward MD, Toney MF (2004) Structural characterization of a pentacene monolayer on an amorphous SiO2 substrate with grazing incidence X-ray diffraction. J Am Chem Soc 126:4084–4085
25. Mannsfeld SCB, Virkar A, Reese C, Toney MF, Bao ZN (2009) Precise structure of pentacene monolayers on amorphous silicon oxide and relation to charge transport. Adv Mater 21:2294
26. Shtein M, Mapel J, Benziger JB, Forrest SR (2002) Effects of film morphology and gate dielectric surface preparation on the electrical characteristics of organic-vapor-phase-deposited pentacene thin-film transistors. Appl Phys Lett 81:268–270
27. Chua LL et al (2005) General observation of n-type field-effect behaviour in organic semiconductors. Nature 434:194–199
28. Jones BA et al (2004) High-mobility air-stable n-type semiconductors with processing versatility: Dicyanoperylene-3, 4 : 9, 10-bis(dicarboximides). Angewandte Chemie-Int Ed 43:6363–6366
29. Jones BA, Facchetti A, Wasielewski MR, Marks TJ (2007) Tuning orbital energetics in arylene diimide semiconductors: materials design for ambient stability of n-type charge transport. J Am Chem Soc 129:15259–15278
30. Chua LL, Ho PKH, Sirringhaus H, Friend RH (2004) High-stability ultrathin spin-on benzocyclobutene gate dielectric for polymer field-effect transistors. App Phys Lett 84:3400–3402
31. Tang ML, Reichardt AD, Wei P, Bao ZN (2009) Correlating carrier type with frontier molecular orbital energy levels in organic thin film transistors of functionalized acene derivatives. J Am Chem Soc 131:5264–5273

# Chapter 5
# Alkylsilane Dielectric Modification Layer: Molecular Length Dependence and the Odd–Even Effect

## 5.1 Introduction

The spin-cast technique discussed in Chap. 4 allowed for the fabrication of alkysilane SAMs on SiO$_2$ of varying chain length. It was determined that longer chain alkylsilanes gave rise to denser films. Furthermore, crystalline monolayers of octadecyltrichlorosilane (which is the most reactive) were formed even without activated hydrolysis using acid or base vapors (see Appendix 4A.1). Shorter chain alkylsilanes the monolayer was as not dense as longer chained ones (determined by GATR-FTIR and GIXD). The ability to create high quality crystalline monolayers at a high throughput on multiple substrates using spin-casting allowed for more rapid investigation into the role of the SAM on the growth of pentacene on different surfaces and corresponding TFT performance.

To study these effects, SAMs with the following alkyl chain length were deposited onto SiO$_2$/Si substrates: C$_6$, C$_7$, C$_8$, C$_{11}$, C$_{12}$, C$_{16}$ and C$_{17}$. The first interesting thing we observed was that at chain lengths of C$_{16}$ or greater the SAM forms a crystalline monolayer. SAMs of C$_{13}$–C$_{15}$ are not commercially available, therefore they were not included in the study. Also since the phase of the SAM was found to be so critical (Chaps. 2–4), for the first part of this chapter some interesting results about the difference between odd and even short chained alkaylsilane SAMs is discussed. The next section describes how often very similar (in terms of measureable experimental parameters) SAMs may give rise to very different pentacene growth and TFT performance. The role of the chain length of the alkylsilane on pentacene TFT performance is then described. The importance of SAM purity and reactivity, and some comments on SAM formation and reproducibility are discussed.

In Chaps. 2 through 4, the role of density of the alkylsilane modified SiO$_2$ on pentacene nucleation, growth, and transistor performance were investigated. In this chapter, for the first time to my knowledge, the *potential* sensitivity of pentacene thin film growth to the number of methylene units in the alkylsilane dielectric

A. Virkar, *Investigating the Nucleation, Growth, and Energy Levels of Organic Semiconductors for High Performance Plastic Electronics*, Springer Theses, DOI: 10.1007/978-1-4419-9704-3_5, © Springer Science+Business Media, LLC 2012

modification monolayer is described. Due to the recent findings about the influence of SAM phase (crystalline vs. amorphous) on pentacene TFT performance, I initially chose to study shorter chain amorphous alkylsilanes with carbon chain lengths of 7, 8, 11, or 12 (past a chain length of 16 the alkylsilane forms a crystalline monolayer as determined by GIXD) [1–3].

Initially, the motivation was to observe if pentacene TFTs were sensitive to whether the underlying SAM consisted of an odd or even number of carbons. For alkanethiols on gold, the terminal methyl group has a different orientation with respect to the rest of the carbon chain and the surface depending on whether the SAM is odd or even. This gives rise to many interesting odd–even effects which have been observed, and will be discussed later in this chapter. For the commonly used even numbered silanes [octyltrimethoxysilane ($C_8$-TS), and dodecyltrimethoxysilane ($C_{12}$-TS)] dielectric modifications pentacene TFTs performed as expected. These transistors showed mobilites ∼ 0.4–0.6 $cm^2V^{-1}s^{-1}$, on/off ratios >$10^6$, and typical field-effect gating. However, for the initial set of studies, pentacene TFTs with odd numbered alkylsilane modifications (heptyltrichlorosilane ($C_7$-TS) and undecyltrichlorsilane ($C_{11}$-TS)) showed no transistor performance (for more than 70 transistors and three different depositions). Using atomic force microscopy (AFM) and grazing incidence X-ray diffraction (GIXD) it was found that on odd numbered alkylsilanes the first few monolayers of pentacene grow highly disordered in-plane leading to an insulating film. On even numbered silanes, on the other hand, the first few deposited pentacene monolayers form a polycrystalline film with grains well organized in-plane. * (*It is important to note that later in the chapter, I will discuss how after more complete tests done on other $C_7$ and $C_{11}$ SAMs where the trichloro group was converted to trimethoxy by quenching in methanol prepared by the LB technique, and by vapor deposition, actually did show TFT behavior with pentacene values showing 0.2–0.5 $cm^2V^{-1}s^{-1}$*).

## 5.2 Initial Experiments Using Odd and Even Length Alkylsilanes as Dielectric Modification Layers in Pentacene TFTs

The odd and even alkylsilane SAMs, with carbon chains consisting of 7, 8, 11 or 12, were characterized by AFM, ellipsometry, water contact angle, GATR-FTIR, and GIXD. The heights of the monolayers were determined from ellipsometry (Table 5.1). For each of the SAMs, the estimated molecular length is longer than the length measured by ellipsometry. This would indicate that the monolayers are fairly disordered.

All the films had nearly identical RMS roughness values—this is important to note since rough films can result in different semiconductor film morphology and in some cases significantly degrade TFT performance [4]. GATR-FTIR was used

**Table 5.1** Properties of the alkylsilane monolayers studied

| Surface treatment | Height [nm] | Extended molecular length [nm] | Contact angle [deg] (SD) | RMS roughness [nm] |
|---|---|---|---|---|
| $C_7$ | 0.8 | 1.1 | 84 (4) | 0.3 |
| $C_8$ | 0.8 | 1.2 | 84 (3) | 0.3 |
| $C_{11}$ | 1.3 | 1.6 | 96 (3) | 0.3 |
| $C_{12}$ | 1.4 | 1.7 | 94 (1) | 0.2 |

**Fig. 5.1** Grazing angle total reflectance-Fourier transform Infrared (GATR-FTIR) spectrum of alkylsilanes studied ($C_7$, $C_8$, $C_{11}$ and $C_{12}$). A slight increase in the absorption area under the longer chain silanes is expected. In all the cases there is no change in the absorption maximum peak position indicating that the order of the alkylsilanes is equivalent

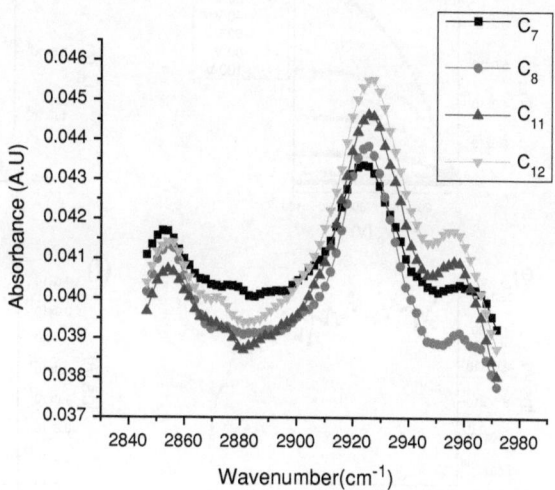

to gauge the ordering and relative densities of the SAMs. The area under the peaks is directly related to the density of the methylene groups in the SAM [5]. The peak area roughly scales with molecular length as expected. For the longer chain SAMs ($C_{11}$ and $C_{12}$) the density of methylene groups is higher since the molecules are taller and also more densely packed (as confirmed by water contact angle measurements) compared to the shorter ($C_7$ and $C_8$) SAMs. The $C_{11}$ and $C_{12}$ SAMs do exhibit a similar degree of ordering and density; the $C_7$ and $C_8$ SAMs are also similarly ordered and dense. All of the SAMs had characteristic $CH_2$ and $CH_3$ stretch mode absorbance maxima at the same wavenumbers. The peak positions at wavenumbers of $\sim 2924$ cm$^{-1}$ and 2855 cm$^{-1}$ indicate that the monolayers are equally disordered and "liquid-like" [5]. Moreover GIXD analysis proved that none of the SAMs were crystalline (no diffraction peak was observed) Fig. 5.1.

The static water contact angles for $C_7$/$C_8$ SAMs and $C_{11}$/$C_{12}$ SAMs (Table 5.1) are very close with slightly lower contact angle for the even SAM in each pair. The small dips in contact angle for even numbered chains has been observed before for alkanethiol/gold systems [6]. As aforementioned, in alkanethiol systems, odd carbon SAMs give rise to a different angle between the terminal methyl group and

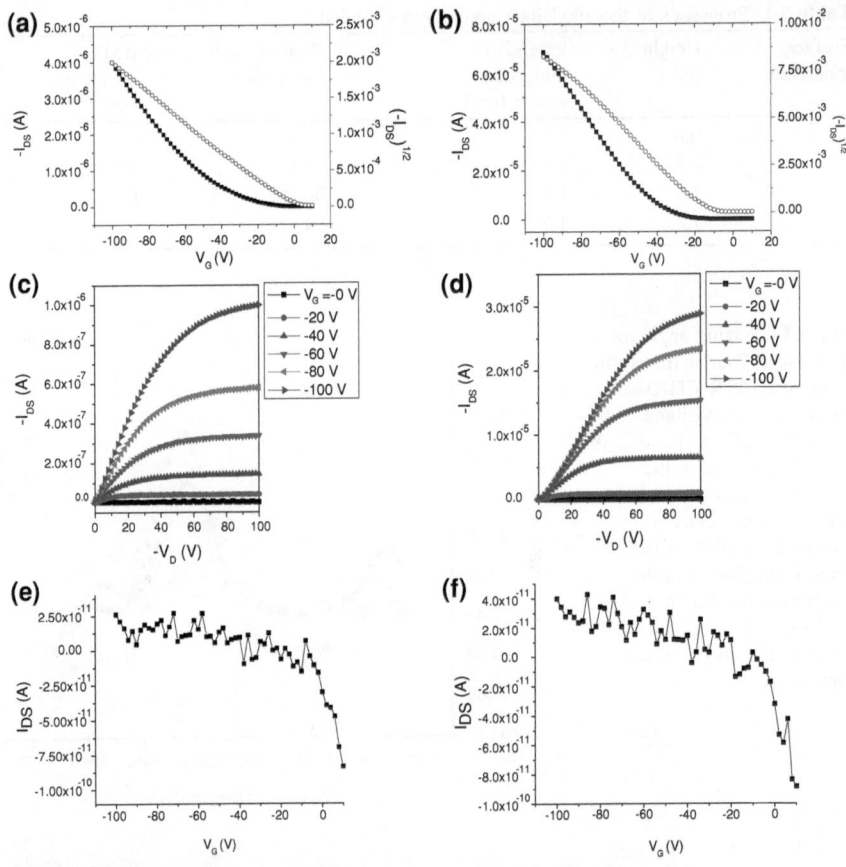

**Fig. 5.2** Pentacene TFT I–V characteristics. **a** typical transfer characteristics for $C_8$ SAM treatment, **b** typical transfer characteristics for $C_{12}$ treatments, **c** typical output characteristics for $C_8$ treatment, **d** typical output characteristics for $C_{12}$ treatment, **e** typical I–V characteristic for $C_7$ treatment, **f** typical I–V characteristics for $C_{11}$ treatment

the head group compared to even numbered SAMs which results in a slightly different surface structure [6].

Initially, I thought that the precise structure of the surface methyl groups in the alkylsilane SAM may also greatly affect the nucleation and growth of pentacene and performance of pentacene TFTs. Figure 5.2 shows typical transfer and output current–voltage (I–V) characteristics for pentacene TFTs on the different SAM modified surfaces. The average electrical performance for over 50 TFTs measured for each SAM dielectric modification are presented in Table 5.2. For even numbered alkyl chains $C_8$ and $C_{12}$, typical pentacene TFT performance is observed. The charge carrier mobilities in the range of $\sim 0.4$–$0.6$ cm$^2$V$^{-1}$s$^{-1}$ and on/off ratios exceeding $10^6$ are consistent for pentacene TFTs with disordered alkylsilane treated dielectrics [1]. The performance on the odd numbered alkyl chains $C_7$ and

**Table 5.2** Pentacene transistor performance on odd and even alkylsilanes

| Surface treatment | $\mu$ [cm$^2$V$^{-1}$S$^{-1}$] | $I_{on}/I_{off}$ | $V_T$ [V] |
|---|---|---|---|
| C$_7$ | ~0 | ~1-2 | NA |
| C$_8$ | 0.4 (0.1) | 10$^6$ | -20 |
| C$_{11}$ | ~0 | 1-2 | NA |
| C$_{12}$ | 0.6 (0.2) | 10$^6$ | -24 |

C$_{11}$ modified dielectric, however, showed no transistor behavior; i.e. there was no increase in source-drain current as a function of increased gate voltage (Fig. 5.2e and 5.2f).

In order to understand the effects of semiconductor microstructure on charge transport in thin film transistors, it is necessary to closely examine the first few monolayers of semiconductor deposited at the semiconductor/dielectric interface [7, 8]. Again I deposited nominally 3 nm thin films of pentacene ($\sim$1.5 monolayers) onto the various SAM surfaces under identical conditions as used for TFT fabrication. Thicker 20 nm and 45 nm films of pentacene were further probed using GIXD discussed later. On the odd numbered SAMs the pentacene growth was highly 3D with most of the islands showing multilayered features. On the even numbered SAMs the pentacene monolayer coverage is also considerably greater than on the odd numbered SAMs, indicating a stronger interaction and wetting tendency on even alkylsilanes (see arguments put forth in Chaps. 2 and 3). It is important to note that the AFM of the 45 nm pentacene thin films deposited on the odd and even alkylsilanes did not show noticeable difference. This reinforces the need to study the morphology of the molecular layers at the dielectric interface.

A much more compelling set of GIXD experiments clearly indicated differences in morphology of the initially grown pentacene on the various surfaces. The GIXD spectrum for pentacene (3 nm, 20 nm and 40 nm thin) films deposited under identical conditions as TFT fabrication on the various SAM treated surfaces is shown in Fig. 5.3. The importance of studying the very thin (3 nm) film has been discussed several times in earlier chapters. Films of 20 nm thickness were deposited since it is approximately at 20 nm where the transition from thin film pentacene to bulk pentacene is typically observed [9]. This transition is related to the relaxation from the strain induced by the substrate. The bulk phase is the thermodynamically most stable molecular packing for a single crystal. The thicker (40 nm) films were probed with GIXD to study if the very different initially deposited film morphology at the dielectric interface propagates the entirety of the film.

The 3 nm pentacene thin film showed pronounced differences between the GIXD spectrum on odd and even alkylsilanes. On C$_8$ and C$_{12}$ (Fig. 5.3b and d) the pentacene diffraction rods are straight, vertical and are parallel with one another. The diffraction spectrum is similar to the reported spectrum for pentacene grown on bother bare silicon oxide (SiO$_2$) and on octadecylsilane modified SiO$_2$ [10]. Both these surfaces (OTS and SiO$_2$) are known be dielectric surfaces onto which pentacene forms a semiconducting channel and working TFTs [1, 2, 9, 11, 12]. The extracted lattice constants for pentacene grown on C$_8$ and C$_{12}$ ($a \approx 5.9$Å,

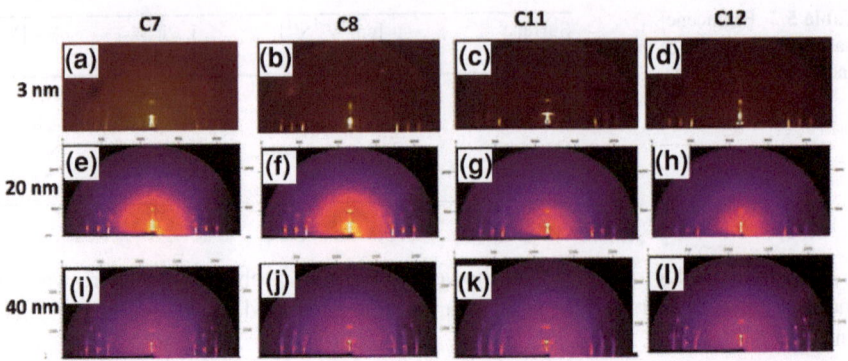

**Fig. 5.3** All pentacene was deposited under identical conditions as used for TFT fabrication; the rate of deposition was 0.3–0.4 Å s$^{-1}$ and the substrate temperature was fixed at 60 °C. GIXD diffractogram: (a–d): pentacene nominally 3 nm deposited on $C_7$, $C_8$, $C_{11}$, and $C_{12}$ respectively. (e–f): pentacene nominally 20 nm deposited on $C_7$, $C_8$, $C_{11}$, and $C_{12}$ respectively. (i–l): pentacene nominally 40 nm deposited on $C_7$, $C_8$, $C_{11}$, and $C_{12}$ respectively

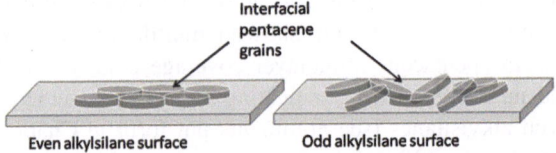

**Fig. 5.4** Schematic showing the morphology of the interfacial pentacene grains directly involved in charge transport, deposited on even and odd numbered alkylsilanes

$b \approx 7.6$ Å and $\gamma \approx 90°$) are consistent with the thin film phase grown on OTS (also even numbered) [13, 14]. The fact that the characteristic $Q_{xy}$ diffraction rods are parallel indicates that the crystalline domains at the semiconductor/dielectric interface are coplanar and that there is little out of plane disorder at the dielectric interface [11]. On the other hand, the diffraction spectrum for 3 nm pentacene on the odd numbered alkylsilanes is very different and shows considerable arching (Fig. 5.3a and c). This arching is indicative of poorly oriented domains at the dielectric interface. The clearly visible diffraction peaks suggest that the film exhibits crystalline domains, but arching indicates out of plane disorder. A completely amorphous sample would give rise to a diffuse ring pattern. The poorly oriented domains in the film result in trap states which severely interrupts charge transport between source and drain electrodes. For thicker pentacene films (20 and 40 nm) several reflections are observed indicative of highly crystalline films regardless of alkylsilane chain length of the underlying SAM (Fig. 5.3). On all of the surfaces, the bulk phase pentacene phase was observed at $Q_{XY} \sim 1.6$ Å$^{-1}$. For pentacene on $C_7$ and $C_{11}$ the arching which was clearly prevalent in the 3 nm films is still evident in the thicker films-but in general the spectra are not significantly different from those on $C_8$ and $C_{12}$ SAMs Fig. 5.4.

## 5.3 Similarities and Differences Between Alkylsilane/SiO2 and Alkanethiol/Gold Systems

From the earlier chapters it was determined that the density of the terminal methyl groups on the surface has a profound effect on pentacene morphology and TFT performance [1–3]. As aforementioned, it is well known that density and structure of the terminal methyl group is different for odd and even numbered alkanethiols on gold [6, 15, 16]. In alkanethiol SAMs the terminal methyl ($CH_3$) group has a different orientation, with respect to the surface bonding thiol group, depending on if the alkyl chain has an odd or even number of carbons (refer to Fig. 5.5 for a schematic showing the structure of even and odd alkanethiol SAMs on gold adapted from Reference 18) [6, 17]. For even numbered alkanehtiols on gold, the transition dipole and the dominant virbrational mode of the terminal methyl group is perpendicular to the surface; for odd numbered alkanethiols the transition dipole and the dominant vibrational mode of the terminal methyl group is more parallel to the surface [17]. It is believed that these differences in active virbational modes and orientation of the transition dipoles give rise to the observed difference in surface properties of odd and even alkanethiols on gold [6, 17]. Odd numbered alkanethiol SAMs have higher surface energy than even numbered ones when they differ by one methylene unit (for example a gold surface with a SAM of $C_{13}H_{27}SH$ has a higher surface energy than one with a $C_{14}H_{29}SH$ SAM) [6]. Also for a consecutive series, the friction coefficient measured by AFM on an odd numbered alkanethiols is also known to be higher than that of even numbered alkanethiols [16]. The odd–even effects in alkanethiol systems are attributed to the inherent ordering and anchoring of the sulfur group onto the crystalline gold surface [6, 17].

In most alkylsilane/$SiO_2$ systems there is no inherent ordered bonding between the silane head group and the amorphous $SiO_2$ surface. However, due to the way the monolayer is prepared, by spin-coating, it is possible that the molecules are locally ordered in very small domains (below the limit of GIXD resolution) and similar effects which occur in alkanethiol systems are taking place [18]. I also measured slightly lower water contact angles (higher surface energy) for odd alkylsilanes compared with even ones.

It is known that the nucleation and growth of molecular semiconductors is highly sensitive to the surface morphology. It was shown previously that crystal growth is extremely sensitive to defects in the SAMs on which they were deposited [19]. In a recent article Biscarini and co-workers showed that the number of methylene units in alkanethiols, which are commonly used to modify the gold electrodes in bottom contact OTFTs, can have significant influences on pentacene TFT performance [15]. Using quantum mechanical calculations they showed that the electronic coupling between pentacene and the alkanethiol was dependent on whether the SAM was odd or even numbered which determined how easily the charges can tunnel from the gold electrode through the alkanethiol and finally into pentacene [15]. The authors found that the charge injection was faster (easier) through even numbered alkanethiol chains compared to odd numbered ones.

**Fig. 5.5** Schematic of odd an even alkanethiol on gold. The arrow indicates the dominant virbational mode and transition dipole

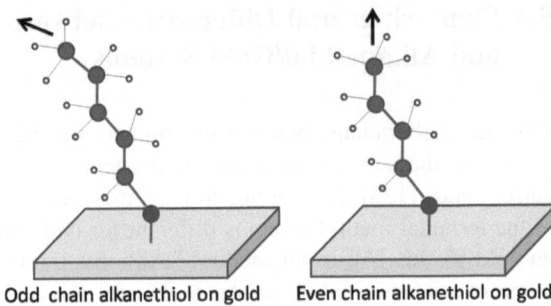

Odd chain alkanethiol on gold     Even chain alkanethiol on gold

In this work a similar effect may be present: the exact electronic structure (dipoles and dominant vibrational stretch modes) of the exposed methyl groups is different depending on whether the chain is odd or even. The nature of these difference may be very complicated to determine, but may be similar to the alkanethiol system where terminal methyl groups have different dominant vibrational and transition dipoles depending on the structure. Maybe this seemingly small difference in odd or even chain length does drastically influence the thin film nucleation and growth behavior of pentacene.

From a technological standpoint, the ability to use different SAMs to pattern pentacene semiconductor growth may allow result in better circuit performance and less cross-talk between devices. *Hermans* and co-workers demonstrated that surfaces with different chemical functionalities can be used to create either discontinuous, or continuous pentacene films [20]. Interestingly in the work presented so far in this chapter, no clear difference in film topology was seen on different silane SAMs. Even though a 40 nm thick pentacene film showed insulting behavior on odd numbered alkylsilanes, presumably due to in-plane disorder at the dielectric interface.

## 5.4 The Role of Alkylsilane Density and Reactivity of Silane Group on the "Odd–Even" Effect

In order to understand more about the odd–even effect, we collaborated with Professor Paulette Clancy's group in the department of Chemical engineering at Cornell Univeristy. The Clancy group works on computer simulations of organic semiconductor nucleation and growth. Recent molecular dynamic simulations conducted by the Clancy group predicted that for alkylsilane systems the pentacene nucleation and growth should become less distinct between odd and even alkylsilanes as the density of the monolayers increase (i.e. the "odd–even" effect should be even more pronounced for $C_7$ and $C_8$ as the SAM densities decreases). Their simulation results showed that the differences in distribution of the angle of the terminal methyl group should increase with decreasing SAM packing density.

They predicted that on low density poorly ordered odd alkylsilane SAMs like $C_7$, pentacene may grow non-coplanar (similar to the phenomenon seen by GIXD). The simulations also seemed to show that on low density $C_7$ surfaces, the differences in the angle of the terminal methyl group gives rise to areas of local energy minima between terminal methyl groups between neighboring $C_7$ molecules. This could potentially lead to non-coplanar and more 3D type pentacene growth. To test this hypothesis Langmuir–Blodgett (LB) films of $C_7$ and $C_8$ were fabricated on $SiO_2$/Si (see Chap. 2). A wide range of densities the SAMs were deposited on $SiO_2$/Si substrates. Pentacene transistors were then fabricated.

To my surprise, on the least dense $C_7$ and $C_8$ SAMs (transferred at surface pressures of 10 or 20 mNm$^{-1}$) fabricated by the LB technique transistor performance was very reasonable 0.2–0.5 cm$^2$V$^{-1}$s$^{-1}$ with an on/off ratio of $10^6$. Recall that from the simulation results it should have been on the least dense $C_7$ where pentacene performance was the poorest. The simulations, however could only calculate diffusivity at a timescale of $\sim 1$ ns, and thus probably does not capture the physics over relevant time scales ($\sim$ on the seconds timescale). There may be batch to batch differences based on the kind of $SiO_2$/Si substrates and even on the quality/type of the silane used. Also, odd chained alkylsilanes may be harder to synthesize of purify, and this may have lead to the observed "odd–even" effect

## 5.5  Role of Alkylsilane Chain Length on Pentacene TFT Performance

The LB experiments (and more vapor and spin-cast experiments) suggested that the differences in pentacene growth and grain orientation initially observed may not be due to the odd–even effect, but perhaps some subtle nature of exactly how the SAM is deposited. What is very interesting is that the SAMs deposited via spin-casting were nearly identical in terms of all measureable experimental parameters (GIXD, FTIR, ellipsometry, and contact angle) but still dramatically affected the pentacene growth. Finally, another potential reason for the differences in the spin-cast and LB films may also be attributed to the high reactivity of the $C_7$ and $C_{11}$ silanes used for spin-casting. $C_7$ and $C_{11}$ SAMs formed by spin-casting were fabricated from the trichlorosilane version of these silanes; the reactivity and sensitivity to moisture of the tricholosilane head group may also have made the SAM formation slightly different than the less reactive (and more reproducible) trimethoxysilane (TMS) versions which are not commercially available. For the LB work described above, the TMS version was synthesized from the trichloro version by quenching with methanol. The tricholorsilane versions are far too reactive with water to be deposited using the LB technique.

Since the "odd–even" effect still remains elusive, and remains an active area of research in our group, the more reliable and commercially available even chain alkylsilanes were also studied. The series of molecules from $C_6$, $C_8$, $C_{12}$, and $C_{16}$

**Fig. 5.6** The charge carrier
mobility of pentacene
transistors as a function of the
length and order of the
dielectric surface
modification monolayer. The
silane density increases as the
chain length increases

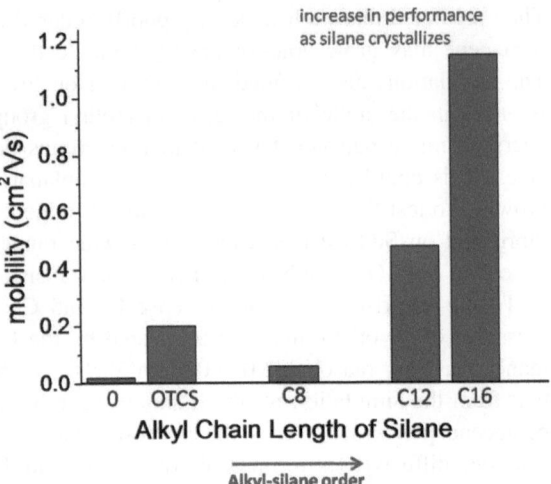

were also deposited on $SiO_2/Si$ using the spin-cast procedure similar to as
described in Chap. 4. Pentacene TFTs were also fabricated. There were several
interesting trends which arose. First, from the spin-cast technique, at chain length
of $C_{16}$, the SAM forms a crystalline monolayer. The GIXD diffractogram of $C_{16}$
clearly showed a peak indicating the SAM was crystalline (see Appendix).

Another interesting, and a nice corollary with the work presented in earlier
chapters about the importance of SAM density on pentacene TFT performance, is
that the pentacene charge carrier mobility increased systematically with alkylsilane SAM chain length; the denser and more ordered the SAM the more 2D the
pentacene growth and the higher the charge carrier mobility Fig. 5.6.

There is a pronounced jump in mobility when the alkylsilane becomes crystalline. Finally, it is important to note that the mobility in the experiments
presented above on the crystalline $C_{16}$-SAM is lower than on crystalline OTS
presented in earlier chapters. Recall the pentacene mobility on crystalline
OTS is $\sim 2$–3 $cm^2V^{-1}s^{-1}$, whereas here the mobility on crystalline $C_{16}$ is
1.2 $cm^2V^{-1}s^{-1}$. This is most likely a reflection of the poorer purity pentacene used
in the experiments presented above. The pentacene purity was only 99.5%,
whereas for the pentacene TFTs used in earlier chapters the purity was greater than
99.99%. As a control, TFTs fabricated using vapor deposited OTS (OTCS) showed
a mobility of $\sim 0.2$ $cm^2V^{-1}s^{-1}$, a factor of 2–3 less than the data presented in
previous chapters where higher purity pentacene was used.

In summary, this chapter contained a detailed study of pentacene growth, and
TFT performance on a series of alkylsilane SAM modified $SiO_2/Si$ surfaces.
Initially very interesting odd–even effects were seen for pentacene on odd and
even short length alkylsilanes. However, after more careful experiments, it was
still difficult to attribute these differences precisely to differences in the number of
carbons in the alkylsilane SAM. The odd even effect seems very sensitive to the
density of the underlying SAM. The exact purity of the SAM, variability in the

SiO$_2$/Si substrates, and potentially some differences in either humidity or SAM reactivity may have given rise to the observed differences in pentacene growth and TFT performance. After the experiments were repeated several times, using a new "fresh" high purity silanes, and using different deposition techniques (LB and vapor deposition) no big odd–even effects were evident. Furthermore, the proposed differences in SAM surface energies and methyl group vibrational modes, which have been reported for alkanethiols on gold, seem to be more complex for alkylsilane SAMs. This is due to the fact that while alkylthiols pack based on sulfur–gold bonds, which determine surface density (and the surface is crystalline), for alkylsilane systems the density is dominated by both bonds between the silane head groups as well as with a poorly defined amorphous SiO$_2$ surface.

The most interesting and useful findings which can be extracted from the work presented is how once again the morphology and growth of the first few nanometers of pentacene at the dielectric surface dominates performance, and how sensitive this growth is to the underlying structure of the dielectric surface. The GIXD confirmed non-coplanar growth and poor grain orientation on some odd numbered alkylsilane SAMs which lead to insulating behavior!

Finally, to complete the work on pentacene on alkylsilanes of varying length, and to learn more about formation of crystalline alkylsilane SAMs from the spin-cast technique developed, heptadecyltrimtheoxysilane (C$_{17}$-TMS) was synthesized (by Song Hyun a visiting chemist). The material was checked for purity and distilled until it was $\sim$99% pure (based on nuclear magnetic resonance NMR). The high purity odd alkylsilane was deposited via the spin-cast technique. GIXD diffractograms showed that C$_{17}$-TMS was crystalline and packed with the same 4.2 Å hexagonal lattice constant as both OTS (C$_{18}$-TMS) and C$_{16}$-TMS (see Appendix). The pentacene TFT performance on the C$_{17}$-TMS was also quite good (1.9 cm$^2$V$^{-1}$s$^{-1}$ averaged over 14 TFTs). This was an interesting finding, since to the best of my knowledge, it was the first odd-chain alkylsilane SAM which has been shown to be crystalline. Fundamental studies about the properties of these different crystalline odd and even alkylsilane SAMs may be of interest to many research groups studying ultrathin films and SAM formation.

## 5.6  Conclusions

In this chapter, the importance of the dielectric/semiconductor interface in organic transistors was further investigated. It was seen that even on similar surfaces (in terms of the experimental properties that can be studied) very different organic semiconductor growth behavior can occur. This chapter contained information which hopefully can instruct those working on organic transistors about some common pitfalls. When I first arrived at what I thought was the "odd–even effect", I was excited and thought it could be a significant finding. However, after thinking more about the alyklsilane SAM formation on SiO$_2$, which is significantly different from the more reproducible and better studied alkanethiol on gold system, and doing some more investigation and discussion with chemists who have expertise in silane chemistry, I realized I needed to do more control experiments. After working with high purity odd and even silanes, and those with similar reactivities, no odd–even effect was observed. In general, it's good advice to perform many control experiments when working with organic transistors since the purity and sensitivity of many organic molecules can by highly variable and getting reproducible results requires good quality materials. There is on-going work (which will be discussed in Chap. 8) being conducted to study more about the alkylsilane SAMs as a surface modification layer in organic transistors. Finally, once pure silanes were used, the data in this chapter follows nicely with the previous chapters. The highest pentacene mobility TFTs were those which were fabricated using crystalline alkylsilane SAMs.

## 5.7 Experimental

### 5.7.1 Materials

Heptyltrichlorosilane, octyltrichlorosilane, undecyltrichlorosilane, and dodecyltrichlorosilane (95% purity), octyltrimethoxysilane (>97% purity) and dodecyltrimethoxysilane (98%) were purchased from Gelest Inc. Device substrates consisted of heavily doped Si wafers with 300 nm of thermally grown silicon oxide having a capacitance per unit area ($Ci$) of 10 nFcm$^{-2}$. Pentacene (99.9%) was purchased from Sigma-Aldrich. The pentacene was sublimed twice prior to fabrication of TFTs with $C_{17}$ silane modification. For ellipsometry and GIXD experiments, silicon wafers with 2–3 nm of native oxide were used. Prior to OTS treatment the wafers were cleaned with piranha (70:30 $H_2SO_4$: $H_2O_2$) for 60 min, washed copiously with DI water, dried with a nitrogen gun and then with ozone (Jetlight UVO-cleaner Model 42–100 V) for 10 min.

### 5.7.2 Deposition of Alkylsilane Monolayers

Alkylsilane solutions (3–4 mM in trichloroethylene) was prepared and filtered (0.2 μm pore size) in a nitrogen glovebox. The solution ($\sim$ 0.5 mL) was dispensed onto the Si/SiO$_2$ substrate until the entire surface was covered. After waiting for 10 s, the spin coater was turned on (3000 RMP) for 10 s. The substrate was then placed into a large bottle (200 mL) along with a small vial (2 mL) of HCl or NH$_4$OH, which catalyzes OTS polymerization. The bottle was then closed and set aside in ambient conditions for 10 h. The substrates were sequentially sonicated in toluene, acetone, isopropanol for 10 min and then dried using a nitrogen gun (99.9% pure). If a residual cloudy polymerized film was observed, a swab-sponge soaked in toluene was used to lightly brush over the surface to remove any unbound-multilayer film.

### 5.7.3 Characterization of SAMs and Pentacene Thin Films

Self-assembled silane monolayers were analyzed using FTIR to study aliphatic stretching modes and to confirm similar molecular packing and density. The grazing angle attenuated total reflectance (GATR) spectrum was obtained using a Nicolet 6700 Fourier Transform Infrared Spectrometer (FTIR) using a germanium crystal. In order to further gauge molecular packing and surface energy, static water contact angle was measured by an Edmund Scientific goniometer. Ellipsometry was used to characterize thickness of the organosilane monolayer. A Sopra Bois-Columbes ellipsometer with a Physike Instrumente laser (He–Ne, $\lambda = 632.8$ nm, angle of incidence of 70°) and detector were used for OTS thickness measurements. Thickness was calculated from $\Psi$ and $\Delta$ values and measured for five areas on the substrate. For data modeling the following input refractive indices ($n_i$) were used : air, $n_0 = 1.0$; alkylsilane, $n_1 = 1.450$; native silicon oxide, $n_2 = 1.460$, silicon, $n_3 = 3.873$, k = -0.016.

The AFM images of organic semiconductor thin films were collected using a Digital Instruments MMAFM-2 scanning probe microscope. Tapping mode AFM was performed on the samples with a silicon tip with a frequency of 300 kHz.

GIXD was performed at the Stanford Synchrotron Radiation Laboratory (SSRL) on beam line 11-3 with a photon energy of 12.73 keV. The angle of incidence was kept fixed at 0.1°.

Following formation of the SAM monolayers, pentacene was deposited at 0.3 Ås$^{-1}$ at a pressure of $\sim 10^{-6}$ Torr to a thickness of 45 nm (as measured by quartz microbalance) onto heated substrates (60 °C). The top-contact transistors were completed by thermal evaporation of gold through a shadow mask defining source and drain electrodes; the channel length was 50 μm and channel width was 1000 μm. It is also important to note that the SAM layers do not undergo noticeable changes as measured by contact angle, ellipsometry, and GATR-FTIR before and after heating to the substrate temperature used for pentacene deposition (60 °C). A Keithley 2400 semiconductor parameter analyzer was used to test p-channel transistors in an ambient. The charge carrier mobility ($\mu$) was calculated by fitting the saturation transfer characteristics using:

$$I_{DS} = \frac{WC}{2L} \mu (V_G - V_T)^2 \tag{5.1}$$

where $I_{DS}$ is the drain current, W (1000 μm) is the channel width, L (50 μm) is the channel length, $C_i$ is the capacitance of the oxide (10 nFcm$^{-2}$), $V_G$ is the gate voltage and $V_T$ is the threshold voltage Fig. 5.7, Fig. 5.8.

# Appendix 5.A.1: Grazing Incidence X-ray Diffractograms of C16 and C17 Alkylsilanes

**Fig. 5.7** GIXD diffractogram of nominally 3 nm pentacene deposited on hexadecylsilane ($C_{16}$) monolayer treated $SiO_2/Si$. The characteristic pentacene Bragg rods are shown. The peak due to the crystalline hexadecylsilane monolayer is highlighted below the large arrow

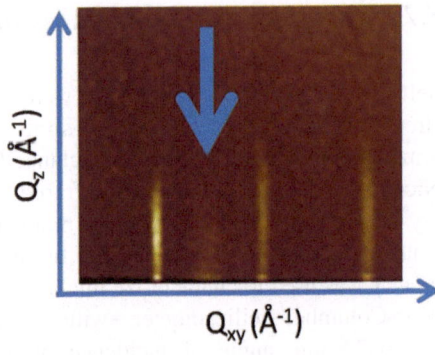

**Fig. 5.8** GIXD diffractogram of the heptadecylsilane ($C_{17}$) monolayer on $SiO_2/Si$. The peak due to the crystalline heptadecylsilane monolayer is highlighted below the large arrow

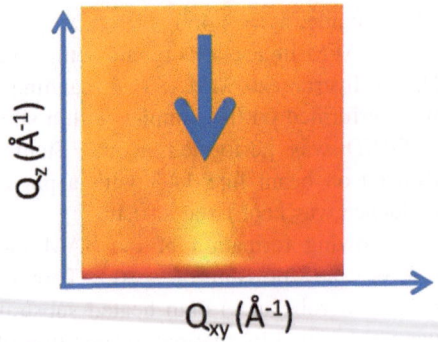

# Appendix 5.A.2: Estimating the Crystallinity of Thin Films Using Grazing Incidence X-ray Diffraction

In organic electronics research it is common to estimate the crystallinity of an organic semiconductor thin film using the Scherrer formula (for more information see [21]. This equation relates the full width half max (FWHM) of a diffraction peak to the coherence length. The coherence length is the size over which a crystal is "correlated" i.e. a larger coherence length is indicative of a larger crystalline domain. From the Scherrer formula, the coherence length (CL) is related to the FWHM of a diffraction peak by:

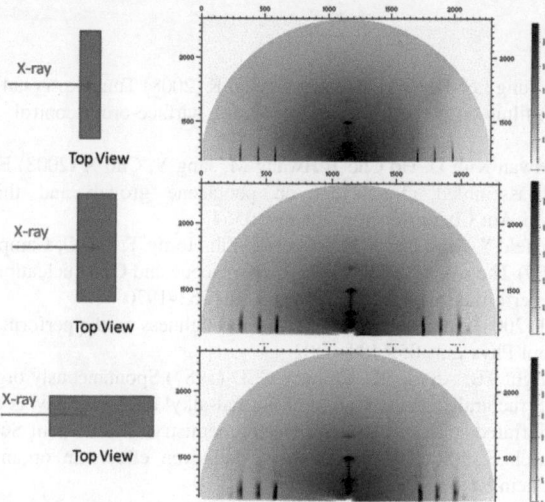

**Fig. 5.9** 20 nm film of pentacene deposited on SiO$_2$/Si. Top: the sample is 0.5 cm in width. Middle: sample is 1.0 cm in width. Bottom sample is 3.0 cm in width. Notice, it looks like the top most sample is the most crystalline (from analysis of Scherrer equation), however all the samples are the same. (In fact the same sample was cut to produce the smaller samples). Thus, the crystalline domains or the polycrystalline thin films are identical, and sample size effects dominate the peak width. It is critical to ensure for thin films, that to even qualitatively compare crystallinity the samples must be identical in size

$$CL = \frac{0.9\lambda}{(FWHM)\cos\theta} \qquad (5.A.1)$$

where $\lambda$ is the wavelength of the X-ray and $\theta$ is the Bragg angle. Analysis of Eq. 5.A.1 indicates that a smaller FWHM (a more narrow peak) is indicative of a more crystalline (larger crystalline domain) films. However, the Scherrer formula which is exact for powder diffraction and a point detector, does not apply to GIXD incidence where the FWHM is detected on a 2D detector. Though many research groups estimate the crystallinity using the Scherrer formula from GIXD, it is incorrect.

For the 2D detector used in these GIXD experiments, every crystallite in a thin film gives rise to a Gaussian diffraction peak, the sum of the peaks is additive and gives rise to the peak observed. The Sherrer formula accurately estimates the coherence length of a single crystalline domain, however in GIXD with a 2D detector, the peak width is dominated by the sample size. Figure 5.9 shows how the same sample of pentacene (20 nm thick) deposited on SiO$_2$/Si shows a different FWHM depending on the sample size. Of course the crystalline grains do not change in size, but blind application of the Scherrer formula would make one think that the samples have different crystallinity. The smaller the sample more crystalline the film appears.

# References

1. Hwan Kim D, Sung Lee H, Yang H, Yang L, Cho K (2008) Tunable crystal nanostructures of pentacene thin films on gate dielectrics possessing surface-order control. Adv Funct Mater 18:1363–1370
2. Sung Lee H, Hwan Kim D, Ho Cho J, Hwang M, Jang Y, Cho K (2008) Effect of the phase states of self-assembled monolayers on pentacene growth and thin-film transistor characteristics. J Am Chem Soc 130:10556–10564
3. Ajay V, Mannsfeld S, Joon Hak O, Michael T, Yih Horng T, Liu G, Campbell S, Robert M, Zhenan B (2009) The role of OTS density on pentacene and C60 nucleation, thin filmgrowth and transistor performance. Adv Funct Mater 19:1962–1970
4. Steudel S et al (2004) Influence of the dielectric roughness on the performance of pentacene transistors. Appl Phys Lett 85:4400–4402
5. Porter MD, Bright TB, Allara DL, Chidsey CED (1987) Spontaneously organized molecular assemblies.4. Structural characterization of normal-alkyl thiol monolayers on gold by optical ellipsometry, infrared-spectroscopy, and electrochemistry. J Am Chem Soc 109:3559–3568
6. Tao F, Bernasek SL (2007) Understanding odd–even effects in organic self-assembled monolayers. Chem Rev 107:1408–1453
7. Park B–N, Seo S, Evans P (2007) Channel formation in single-monolayer pentacene thin film transistors. J Phys D Appl Phys 40:3506–3511
8. Dinelli F et al (2004) Spatially correlated charge transport in organic thin film transistors. Phys Rev Lett 92:116802
9. Ruiz R et al (2004) Pentacene thin film growth. Chem Mater 16:4497–4508
10. Fritz SE, Martin SM, Frisbie CD, Ward MD, Toney MF (2004) Structural characterization of a pentacene monolayer on an amorphous $SiO_2$ substrate with grazing incidence X-ray diffraction. J Am Chem Soc 126:4084–4085
11. Bao Z, Locklin J (2007) Organic field effect transistors. CRC Press Taylor and Francis Group, Boca Raton
12. Halik M et al (2004) Low-voltage organic transistors with an amorphous molecular gate dielectric. Nat 431:963–966
13. Ruiz R et al (2004) Structure of pentacene thin films. Appl Phys Lett 85:4926–4928
14. Yang HC et al (2005) Conducting AFM and 2D GIXD studies on pentacene thin films. J Am Chem Soc 127:11542–11543
15. Pablo Stoliar RK, Massi M, Annibale P, Albonetti C, de Leeuw DM, Biscarini F (2007) Charge injection across self-assembly monolayers in organic field-effect transistors: odd–even effects. J Am Chem Soc 129:6477–6484
16. Mikulski PT, Herman LA, Harrison JA (2005) Odd and even model self-assembled monolayers: links between friction and structure. Langmuir 21:12197–12206
17. Ulman A (1991) An introduction to ultrathin organic films from Langmuir–Blodgett to self assembly. Academic Press, San Diego
18. Nie H-Y, Walzak MJ, McIntyre NS (2006) Delivering octadecylphosphonic acid self-assembled monolayers on a Si wafer and other oxide surfaces. J Phys Chem B 110:21101–21108
19. Aizenberg J, Black AJ, Whitesides GM (1999) Control of crystal nucleation by patterned self-assembled monolayers. Nat 398:495–498
20. Steudel S, Janssen D, Verlaak S, Genoe J, Heremans P (2004) Patterned growth of pentacene. Appl Phys Lett 85:5550
21. Mannsfeld SCB, Virkar A, Reese C, Toney MF, Bao ZN (2009) Precise structure of pentacene monolayers on amorphous silicon oxide and relation to charge transport. Adv Mater 21:2294

# Chapter 6
# Low-Voltage Monolayer Pentacene Transistors Fabricated on Ultrathin Crystalline Self-Assembled Monolayer Based Dielectric

## 6.1 Introduction

In this final chapter on organic transistors, two of the issues limiting the fabrication of low-cost, high performance transistors are addressed by using concepts presented in earlier chapters. The first major hurdle is associated with wasting materials during the deposition processes. Using a thinner layer of the semiconductor (much thinner than 40 nm usually used) allows for a lower material cost. In this work, the two-dimensional growth of pentacene on crystalline OTS was utilized to fabricate monolayer thin transistors (M-TFTs), wherein the active semiconducting channel is only one monolayer in thickness. The second major issue is the very high operating voltages associated with a thick dielectric ($\sim$100 V). To resolve this issue M-TFTs were fabricated on an ultrathin (OTS) gate-dielectric; this allows for operating voltages of $\sim$3 V or less.

Pentacene monolayer transistors are fabricated on a crystalline monolayer of octadecylsilane (OTS) modified $SiO_2$. Mobilities as high as 0.12 cm$^2$ V$^{-1}$ s$^{-1}$, and on/off ratios $>10^5$ were demonstrated. Also ultra-low voltage monolayer pentacene TFTs were fabricated on a crystalline OTS monolayer dielectric on native silicon oxide. The densely packed OTS served as a fairly good dielectric layer, and monolayer pentacene TFTs showed mobilities of 0.04 cm$^2$ V$^{-1}$ s$^{-1}$. This represents (to the best of my knowledge) the first known demonstration of a monolayer OTFT on an ultrathin gate dielectric.

As discussed in earlier chapters, for a commonly used bottom gated structure, the applied gate field confines the induced mobile charge to within the first few nanometers of semiconductor at the dielectric/semiconductor interface, so a contiguous monolayer of semiconductor can serve as the active layer [1, 2]. There have been several recent breakthroughs in the field of organic M-TFTs. Smits and co-workers created M-TFTs using a molecule which was composed of a thiophene based semiconducting moiety covalently attached to an aliphatic silane tail. The M-TFT could easily be formed from solution with the terminal silane group

A. Virkar, *Investigating the Nucleation, Growth, and Energy Levels of Organic Semiconductors for High Performance Plastic Electronics*, Springer Theses, DOI: 10.1007/978-1-4419-9704-3_6, © Springer Science+Business Media, LLC 2012

binding to the silicon oxide ($SiO_2$) dielectric [3]. Mobilities as high as 0.04 $cm^2$ $V^{-1}$ $s^{-1}$were achieved. This approach is quite elegant, since the M-TFTs were created by simply allowing the semiconductor molecule to self assemble from solution. M-TFTs were fabricated reproducibly over large areas (4 inch wafers) [3].

Asadi fabricated pentacene M-TFTs on unmodified $SiO_2$ with bottom contact metal electrodes treated with thiols that improved pentacene growth and demonstrated mobilities as high as 0.05 $cm^2$ $V^{-1}$ $s^{-1}$ [4]. Finally, Huang and co-workers showed that chemical sensors based on mono or bilayer TFTs on unmodified $SiO_2$ are extremely sensitive since the molecules sensing the analyte are also directly involved in charge transport [5]. The field of OTFT sensors is an exciting and promising one, and its success may be dictated by the ability to create high mobility, low-cost, high sensitivity M-TFTs.

However, previous work on M-TFTs had been based on depositing a monolayer on bare $SiO_2$ [3–5]. The surface energy of $SiO_2$ is much higher than that of organic semiconductors, so the semiconductor grows two dimensionally (2D) on bare $SiO_2$ (see Chap. 3) [6]. There are several problems associated with the semiconductor/ $SiO_2$ interface, including trapped water and dendritic semiconductor growth, which have been described in Chaps. 1 and 2 [7, 8]. For future commercial organic electronic applications it is more likely the dielectric will be organic, and thus creating M-TFTs on organic surfaces is desirable [9, 10].

## 6.2 Pentacene Monolayer Transistors on Crystalline OTS Modified $SiO_2$

To fabricate pentacene M-TFTs the 2D growth of pentacene on crystalline OTS was exploited. Initially, to fabricate the M-TFTs, highly doped silicon wafers (n++) with 300 nm of thermally grown oxide served as the substrate. The substrates were cleaned with pirhanna (7:30 $H_2SO_4/H_2O_2$) and then OTS by the spin-casting technique described in Chap. 4 [11]. Pentacene (nominally 2–3 nm in thickness as monitored by a quartz microbalance) was then deposited onto the OTS treated $SiO_2/Si$ substrates at a pressure of $10^{-6}$ torr, and a rate of 0.3–0.4 Å $s^{-1}$; the substrate temperature was held at 60 °C. Finally, gold was thermally evaporated (at 0.5 Å $s^{-1}$ to a film thickness of 40 nm) through a shadow mask to define source and drain electrodes (channel width: 1,000 μm; channel length: 50 μm). The relatively large distance between the electrodes required highly 2D pentacene growth and a contiguous layer.

The atomic force micrograph of the pentacene channel on the crystalline-OTS (see Chap. 4) treated substrate is presented in Fig. 6.1. On the crystalline OTS the growth of pentacene is 2D over a large area (image area is 400 μm$^2$). The pentacene wets the surface forming a continuous 2D monolayer sheet. The height difference between the monolayer and underlying OTS was measured to be about 1.8 nm, which is close to the molecular length of pentacene ($\sim$ 1.4–1.5 nm) [12].

**Fig. 6.1  a** Atomic force micrograph of pentacene nominally 2.6 nm on crystalline OTS on SiO$_2$/Si substrate **b** the corresponding line profile

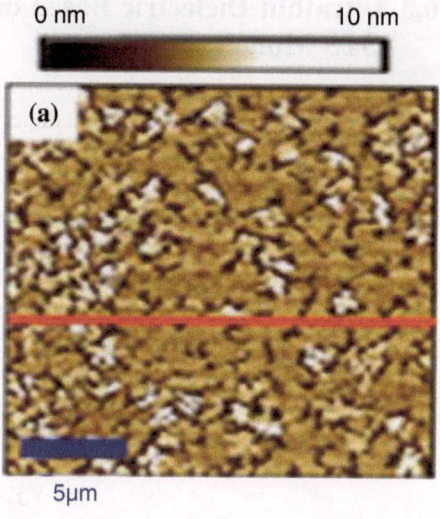

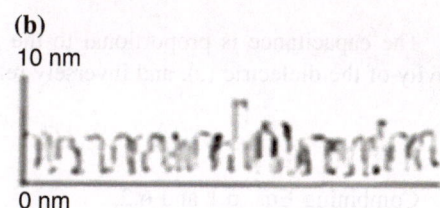

**Fig. 6.2** Typical pentacene M-TFT transfer I–V plot. The *dashed line* shows the slope used to calculate mobility. The mobility, on/off ratio and threshold voltage are presented as an *inset*

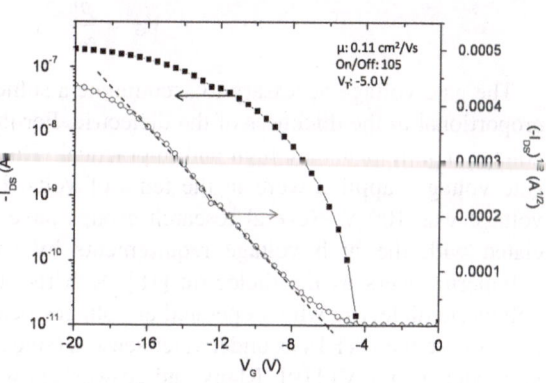

On the crystalline OTS layer, the monolayer of pentacene was sufficient to function as the active layer in a TFT. Figure 6.2 shows a typical transfer current–voltage (I–V) curve for the M-TFT. The mobility averaged over ten devices was 0.09 cm$^2$ V$^{-1}$ s$^{-1}$ ($\pm$0.02) with a maximum value of 0.12 cm$^2$ V$^{-1}$s$^{-1}$, the threshold voltage was $-5$ (3) and the on/off ratio was $\sim 10^5$. The values for mobility here are substantially higher than previously reported for those on bare SiO$_2$ [3, 13]. The improvement is probably due to the removal of hydroxyl traps and improved pentacene growth on OTS compared to bare SiO$_2$ [14].

## 6.3 Ultrathin Dielectric Based on a Crystalline OTS Monolayer

The benefits of crystalline OTS as a dielectric surface modification layer have been well established over the past several chapters. However in all the previous chapters, all the OTS monolayers used for TFT fabrication have been deposited on $SiO_2$/Si substrates; the $SiO_2$ on these substrates is a thermally grown oxide which is 300 nm in thickness. Typically thick layers of oxide are necessary to provide a barrier against high leakage current [15, 16]. These substrates are very commonly in research and are produced by the companies fabricating silicon wafers. The 300 nm thick layer requires very high operating voltages to generate the electric fields necessary to accumulate mobile charges in the channel. The charge density ($\rho$) in the channel induced by the applied gate voltage ($V_G$), is related to the capacitance ($C$) of the dielectric by:

$$V_G = \frac{\rho}{C} \tag{6.1}$$

The capacitance is proportional to the permittivity of free space ($\varepsilon_o$), permittivity of the dielectric ($\varepsilon$), and inversely related to the dielectric thickness (t).[17].

$$C = \frac{\varepsilon_0 \varepsilon}{t} \tag{6.2}$$

Combining Eqs. 6.1 and 6.2,

$$V_G = \frac{\rho t}{\varepsilon_0 \varepsilon} \tag{6.3}$$

The gate voltage necessary to accumulate a sufficient charge density is directly proportional to the thickness of the dielectric. For the 300 nm thick dielectrics the voltage requirements are high and impractical. This is why in earlier chapters the gate voltages applied were in the ten's of volts. Usually the maximum applied voltage was 100 V. Several research groups have addressed the problems associated with the high voltage requirements by trying to use thin (10–20 nm) polymeric layers as the dielectric [18]. Roberts et al. developed a cross-linked polymeric dielectric which operated at voltages below 1 V [19–21]. This allowed them to use the OTFTs as underwater sensors (since the hydrolysis of water occurs at greater than 1 V) [19]. Klauk and co-workers were pioneers in using ultrathin self-assembled monolayers of alkyl phosphonic acid (such as octadecylphosphonic acid (OPA)) on very thin ~2–3 nm aluminum oxide ($Al_2O_3$) layers. Using a gate dielectric which was composed of a SAM of OPA on $Al_2O_3$ voltages lower than 3 V were used to operate high performance pentacene TFTs [17].

Though $Al/Al_2O_3$ are attractive as substrates for organic electronics, the $Al_2O_3$ is often fabricated using atomic layer deposition (ALD) to make an ultrasmooth surface. ALD may not be amenable for large area or cost effect production. Extending from Klauk and co-workers, it was hypothesized that the crystalline OTS layer on $SiO_2$/Si could also serve as a gate dielectric in OTFTs. It should be noted, that heavily doped silicon wafers without 300 nm thermal oxide are also readily available. However, when the silicon is exposed to ambient conditions the top-most atomic layers oxidize to form a native oxide which is typically 2–3 nm in thickness and also quite smooth (typically same smoothness as underlying Si $\sim 0.1$–$0.3$ nm). Klauk and co-workers also showed that for OPA and $Al_2O_3$, the SAM dielectric works well since the OPA forms a dense and highly ordered monolayer (though they have yet to characterize if the OPA is actually crystalline, from the contact angle and ellipsometric data I am quite confident the monolayer is crystalline). They also described that their previous attempts using OTS on native $SiO_2$ failed [16]. The OTS was not well ordered and dense enough and leakage currents were too high to operate a transistor. However, since the crystalline OTS layer we developed is much more dense (about 1.45 times) than typically OTS films made previously by other groups, it could serve as a monolayer dielectric on native $SiO_2$ [11].

The crystalline OTS layer was deposited on a heavily doped Si substrate with 2–3 nm of native $SiO_2$ using the spin-cast technique described in Chap. 3. To test initially if the OTS could serve as a dielectric, 40 nm thick pentacene transistors were fabricated. The transistors worked quite well (under identical conditions as described in previous chapters). Unfortunately, measuring the exact capacitance of the OTS monolayer is difficult. The capacitance is necessary for accurate determination of the charge carrier mobility (see Eq. 1.2). While most research groups deposit metals like gold onto the SAM via thermal evaporation to measure capacitance, this has been shown to be inaccurate [22]. The hot and heavy gold atoms typically penetrate the SAM layer. In fact, making contacts to molecular junctions has been a difficult task and a major research effort in the field of molecular electronics [22]. Realizing the inherent problems in accurately measuring capacitance via thermal evaporation of gold onto the OTS, another method was explored. The gold was deposited onto an elastomeric (polydimethylsiloxane PDMS) substrate. Then the PDMS with patterned gold was placed in contact with the OTS layer. However, reproducible and reliable capacitance measurements were difficult to make since the measured capacitance changed if too much or too little pressure was applied to OTS by the gold covered PDMS. Also as the PDMS flexed and bent the gold film cracked which also lead to inaccuracies in the measured capacitance. Only recently, Akkerman and co-workers demonstrated a robust and reliable method to probe conductivities and capacitances of SAMs. They used a conductive polymer as a buffer layer to prevent damage to the SAM layer during thermal evaporation of gold [22]. For more details see Ref. [22]. Failures to accurately measure capacitance lead me to assume that the capacitance of the OTS monolayer was similar to the OPA monolayers which have been extensively characterized [17]. Both SAMs are eighteen carbons in length and of comparable density. The measured thickness of native $SiO_2$ was 2.4 nm and OTS was 2.1 nm. The total capacitance of the 2.4 nm $SiO_2$ plus the OTS can be calculated by:

$$C_{\text{tot}} = \frac{1}{\frac{1}{C_{\text{OTS}}} + \frac{1}{C_{SiO_2}}} \qquad (6.4)$$

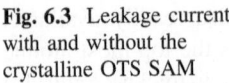

**Fig. 6.3** Leakage current
with and without the
crystalline OTS SAM

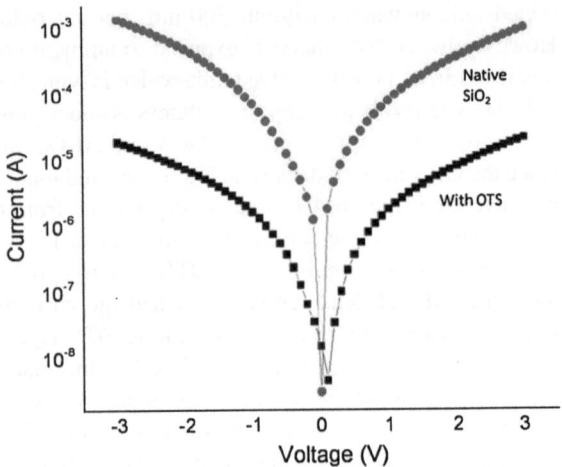

For the OTS, the values calculated by Klauk et al. were used, and I have assumed a value of 1.05 $\mu F\ cm^{-2}$ for $C_{OTS}$ (the capacitance of OTS). $C_{SiO2}$ was calculated to be 1.4 $\mu F\ cm^{-2}$ [17]. The total capacitance for the OTS/SiO$_2$ hybrid dielectric was then calculated using Eq. 6.4 and was found to be 0.61 $\mu F\ cm^{-2}$ (a value fairly close to the OPA/Al$_2$O$_3$ system 0.7 $\mu Fcm^{-2}$ ) [17].

As aforementioned, the major problem with using low density OTS SAMs as dielectric was very high leakage currents. In order to test the leakage current with and without the crystalline OTS SAM, circular gold electrodes (diameters between 501,000 $\mu m$) were deposited onto the crystalline OTS treated native SiO$_2$ substrate, as well as the bare SiO$_2$ substrate. The leakage current is shown in Fig. 6.3.

The leakage current is two orders of magnitude lower when the crystalline OTS layer is used as the dielectric. Also, it should be noted that this is an underestimate of the differences in the leakage current since again the gold deposited penetrates the OTS layer causing damage and potential for shorts. In the completed top contact transistor, the semiconductor acts a buffer layer to prevent degradation to the OTS layer. Using the estimated value for the capacitance, the 40 nm pentacene TFTs had mobilities of 1.8 cm$^2$ V$^{-1}$ s$^{-1}$ (0.3). The on/off ratio was consistently above 10$^6$, and the average threshold voltage was $-12$ V ($-6$) (averaged over ten transistors). The maximum gate voltage applied was $-3$ V.

## 6.4 Low Power Monolayer Thick Pentacene Transistors

The ability to fabricate pentacene TFTs with good performances on the ultra thin dielectrics help further the utility of the crystalline OTS SAM in plastic electronic devices. Finally, the two concepts from this chapter: a monolayer pentacene transistor and an ultrathin monolayer dielectric using crystalline OTS were combined. To fabricate the transistors, a monolayer of pentacene (nominally 2.2 nm) was deposited on the crystalline OTS modified native $SiO_2$ substrate. The substrate temperature was held at 60 °C (which from earlier work was determined to be the optimal temperature for our system). Pentacene was deposited at a rate on 0.3–0.4 Å $s^{-1}$. The monolayer pentacene TFTs on the ultrathin dielectrics performed quite well. The average charge carrier mobility was 0.04 cm$^2$ V$^{-1}$ s$^{-1}$ (0.01), and the on/off ratios was $10^4$ (averaged of five devices). The major problem with these devices was the stability of pentacene. For 40 nm thick films there is a greater barrier between the ambient (oxygen and water both known to degrade pentacene TFT performance) and the active channel. The monolayer of pentacene in the M-TFT was very susceptible to degradation, and the TFT could only be operated about ten times in ambient (about 20 min) prior to device failure. The failure may be avoided by using an encapsulation layer [15].

## 6.5 Conclusions

In conclusion, by utilizing the highly 2D growth mode and strong interaction energy of pentacene with crystalline OTS, monolayer pentacene TFTs were fabricated on organic surfaces. Moreover, using the dense crystalline OTS layer, ultrathin dielectrics could be fabricated and the TFT maximum operating voltage was reduced from −100 to −3 V. Combining both the 2D growth of pentacene on crystalline OTS, and the ability for crystalline OTS to serve as a dielectric, the first ultralow power monolayer organic transistors were fabricated. If the devices can be encapsulated properly they may be very attractive as building blocks for organic circuitry since both low operating voltages can be used, and the amount of organic semiconductor needed is considerably less which ultimately reduces the time and cost required to fabricate an OTFT.

## References

1. Dinelli F, et al (2004) Spatially correlated charge transport in organic thin film transistors. Phys Rev Lett 92(116802):1–4
2. Dodabalapur A, Torsi L, Katz HE (1995) Organic transistors-2-dimensional transport and improved electrical characteristics. Science 268:270–271
3. Smits ECP et al (2008) Bottom-up organic integrated circuits. Nature 455:956–959
4. Asadi K, Wu Y, Gholamrezaie F, Rudolf P, Blom PWM (2009) Single-layer pentacene field-effect transistors using electrodes modified with self-assembled monolayers. Adv Mater 21:4109–4114
5. Huang J, Sun J, Katz HE (2008) Monolayer-dimensional 5,5′-Bis(4-hexylphenyl)-2,2′-bithiophene transistors and chemically responsive heterostructures. Adv Mater 20:2567
6. Yoshikawa G et al (2007) Spontaneous aggregation of pentacene molecules and its influence on field effect mobility. Appl Phys Lett 90(251906):1–3
7. Chua LL et al (2005) General observation of n-type field-effect behaviour in organic semiconductors. Nature 434:194–199

 8. Shankar K, Jackson TN (2004) Morphology and electrical transport in pentacene films on silylated oxide surfaces. J Mater Res 19:2003–2007
 9. Puigdollers J et al (2004) Pentacene thin-film transistors with polymeric gate dielectric. Org Electron 5:67–71
10. Roberts ME, Mannsfeld SCB, Stoltenberg RM, Bao ZN (2009) Flexible, plastic transistor-based chemical sensors. Org Electron 10:377–383
11. Ito Y et al (2009) Crystalline ultrasmooth self-assembled monolayers of alkylsilanes for organic field-effect transistors. J Am Chem Soc 131:9396–9404
12. Mannsfeld SCB, Virkar A, Reese C, Toney MF, Bao ZN (2009) Precise structure of pentacene monolayers on amorphous silicon oxide and relation to charge transport. Adv Mater 21:2294–2298
13. Asadi K, Gholamrezaie F, Smits ECP, Blom PWM, De Boer B (2007) Manipulation of charge carrier injection into organic field-effect transistors by self-assembled monolayers of alkanethiols. J Mater Chem 17:1947–1953
14. Wo ST et al (2006) Structure of a pentacene monolayer deposited on SiO2: role of trapped interfacial water. J Appl Phys 100
15. Bao Z, Locklin J (2007) Organic field effect transistors. CRC Press Taylor and Francis Group, Boca Raton
16. Halik M et al (2004) Low-voltage organic transistors with an amorphous molecular gate dielectric. Nature 431:963–966
17. Klauk H, Zschieschang U, Pflaum J, Halik M (2007) Ultralow-power organic complementary circuits. Nature 445:745–748
18. Klauk H et al (2002) High-mobility polymer gate dielectric pentacene thin film transistors. J Appl Phys 92:5259–5263
19. Roberts ME et al (2008) Water-stable organic transistors and their application in chemical and biological sensors. In: Proceedings of the National Academy of Sciences of the United States of America vol 105, pp 12134–12139
20. Roberts ME et al (2009) Cross-linked polymer gate dielectric films for low-voltage organic transistors. Chem Mater 21:2292–2299
21. Roberts ME, Sokolov AN, Bao ZN (2009) Material and device considerations for organic thin-film transistor sensors. J Mater Chem 19:3351–3363
22. Akkerman HB, Blom PWM, de Leeuw DM, de Boer B (2006) Towards molecular electronics with large-area molecular junctions. Nature 441:69–72

# Chapter 7
# Highly Conductivity and Transparent Carbon-Nanotube and Organic Semiconductor Hybrid Films: Exploiting Organic Semiconductor Energy Levels and Growth Mode

## 7.1 Introduction

The previous chapters have focused on understanding and controlling organic semiconductor growth for high performance organic transistors. In this chapter, the lessons learned from studying organic semiconductor nucleation and growth for transistors are applied to improve the conductivity of carbon nanotube (CNT) networks for transparent electrode applications. A carbon nanotube network (CNTnw) is a mono (or multi) layer of nanotubes deposited on a substrate (schematically shown in Fig. 7.1).

CNTnws are typically prepared by dispersion of the CNTs in a solvent. The dispersion is then cast onto a substrate by a variety of techniques including dip coating, spin-coating, spray coating, or a variety of other solution coating techniques. The solvent is then dried/evaporated to form a CNTnw.

## 7.2 Transparent Electrodes

One of the major applications envisioned for CNTnws are transparent electrodes. Transparent electrodes (TEs) are vital components in solar cells, flat panel displays, and touch screens, wherein materials with high electrical conductivity and low optical absorption are required [1]. However, optical transparency and conductivity are inversely related, and thus represent an interesting and challenging material science problem. Currently, the material most widely used for transparent electrodes is indium tin oxide (ITO), though other doped metal oxides are also promising candidates (like aluminium doped zinc oxide, fluorine tin oxide). Electronically, metals oxide have a large bandgap and are optically transparent (in the visible region of the spectrum), however by doping, the impurity states offers free charges for charge transport. There are, however, many problems associated with

A. Virkar, *Investigating the Nucleation, Growth, and Energy Levels of Organic Semiconductors for High Performance Plastic Electronics*, Springer Theses, DOI: 10.1007/978-1-4419-9704-3_7, © Springer Science+Business Media, LLC 2012

**Fig. 7.1** A schematic of a
carbon nanotube network.
The *lighter tubes* represent
semiconducting CNTs, and
the *darker tubes* represent
semiconducting CNTs

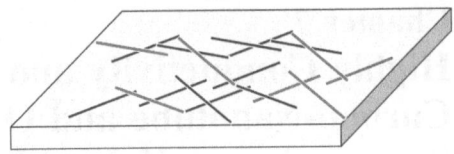

ITO (and other metal oxides) including the price and scarcity of indium, high
processing temperatures and brittleness [1]. For many thin film solar technologies,
up to 25% of the cell cost may be due to the metal oxide transparent electrode.
Furthermore, the high processing temperatures ($\sim$ 100–500 °C) render ITO (and
other metal oxides) incompatible with flexible substrates.

Aside from CNTs, there are a few other materials being considered as
replacements for doped metal oxides for TE applications. These include metal
nanowire networks and graphene [2–4]. The strategy for nanowire meshes is very
similar to CNTnws, i.e. highly conductive silver nanowires are deposited as a thin
film from solution to form a conductive network [3]. Also, similar to CNTs, even
though the silver nanowires themselves are highly absorptive, thin meshes can be
very transparent since the interstitial areas between nanowires still allow the vast
majority of light to be transmitted [3, 4]. Graphene is another interesting candidate
since it has an incredible sheet conductivity. Currently there are still challenges in
fabricating inexpensive high quality graphene sheets over large areas [2]. The two
primary metrics used to characterize the performance of a TE are transparency and
sheet resistance ($R_s$), which is the inverse of sheet conductivity. Of course since the
transparency is a function of the incident wavelength, by convention, 550 nm has
been chosen since it is in the middle of the visible spectrum range and is an
important wavelength to consider for solar, touch screen and display technologies
[1]. The sheet resistance is simply the three dimensional resistance normalized to
the thickness of the conductor. Thus for thin films or coatings it gives the two
dimensional resistance but is dimensionless in area and thus in units of $\Omega\square^{-1}$,
where the $\square$ represents the square dimensions (but is unit-less). Depending on the
application, the trade off between sheet resistance and transparency is managed. For
example, for solar cells the $R_s$ should be as low as possible, and transparency should
be as high as possible. This allows for the most light to enter the active semiconducting layer and also for the maximum power extraction. ITO used in solar cells
has a $R_s$ of roughly 10–40 $\Omega\square^{-1}$ at 80–90% transparency (at 550 nm) [1]. However, for touch screens the desirable $R_s$ is roughly 250 $\Omega\square^{-1}$ at >93% transparency.

## 7.3  Carbon Nanotube Based Transparent Electrodes

CNTnws are attractive as TEs in the aforementioned applications because CNTs
can have exceptional conductivities (the mobility of a single tube can be 100,000
$cm^2 \ V^{-1} \ s^{-1}$), mechanical integrity, stability, and can be processed inexpensively

from solution onto flexible substrates. Major strides in fabricating conducting films of CNTnws from solution have been reported, but the resulting films are still too resistive, absorptive, or rough (thick) and the metrics are still far poorer than ITO [5–7]. Progress in fabricating CNTnw TEs has been limited due to two major factors. Firstly, as aforementioned in the introductory chapter, the CNTnw contains a mixture of semiconducting and metallic CNTs based on their chirality (the way the graphene sheet is wrapped up). A good approximation is that 2/3 of a CNTnw is composed of semiconducting tubes, and 1/3 metallic [8–10]. To further complicate matters, the bandgap and molecular orbital energies for CNTs are related to the size (diameter) and deformation of the tube, and by defects. The semiconducting tubes are less conductive than the metallic tube, and thus impede the performance of CNTnw TEs. Aside from these "fundamental" limitations with CNTnws, there are also practical issues. Firstly, due to their strong intermolecular interactions, CNTs are typically not soluble in high concentrations in most solvent [11]. Also, since junction resistances dominate thin film conductivity, longer tubes are also desirable, but again they are even harder to disperse than shorter tubes [10].

Many researchers have processed the CNTs with surfactants or polymers in order to improve the solubility of the CNTs, but both lower the overall conductivity of the network [7, 9]. A considerable amount of work had addressed some of the CNT processing issues, but, the performance of CNTnw TEs are still considerably poorer than ITO, with the best doped CNTnw $R_s$ values around $\sim 80\ \Omega\square^{-1}$ at 80% transparency, or $1\ k\Omega\square^{-1}$ at 93% transparency (Fig. 7.2) [9]. Furthermore, these "champion" values were obtained by doping the CNTnw using thionyl chloride or iodine. Both thionyl chloride and iodine are unstable, and the doping is temporary due their high vapour pressure; moreover, thionyl chloride is toxic and can react with other materials in a device making efforts to encapsulate even more challenging [9].

In this chapter a permanent "nanosolder or nanoglue" approach was utilized whereby fullerenes are deposited onto very thin, partially aligned CNTnws to greatly reduce tube–tube junction resistance. Subsequently, thicker films of highly electronegative fullerenes are deposited to provide permanent and stable doping of the semiconducting tube resulting in among the highest performing and most stable CNT based TE.

## 7.4   Fabrication of Hybrid CNT Based Transparent Electrode

The first step towards making a high performance CNTnw base TE is the fabrication of uniform CNTnws with controlled density, alignment, and degree of bundling. Recent theoretical work has shown that an ideal monolayer CNTnw, with minimal resistances, and all metallic tubes, could produce $R_s$ of $1\ \Omega\square^{-1}$ at >95% transparency [10]. For this work we chose to concentrate on optimizing a monolayer (or bilayer) of CNTs to ensure high transparency. Monolayer thin CNTnws were formed from a solution of CNTs were dispersed (stably in

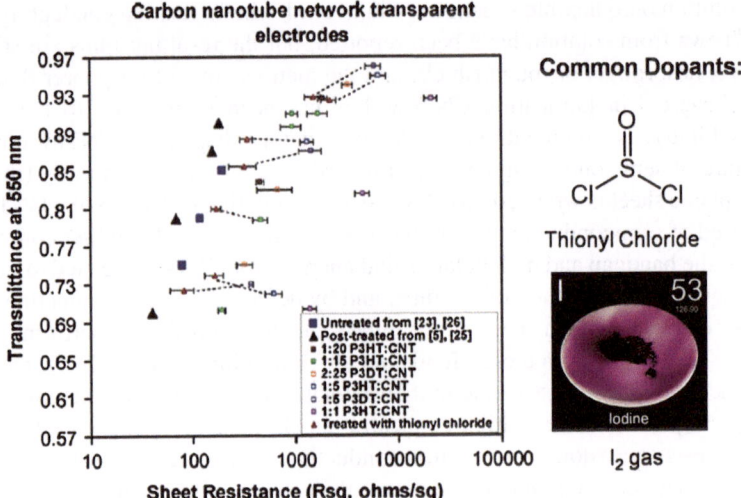

**Fig. 7.2** A compilation of the values published in literature of carbon nanotubes based transparent electrodes. The improvements in sheet resistances by using one of the unstable dopants are shown by the *dashed line*. (From Ref. [9])

concentrations up to 500 µg $ml^{-1}$) in *N*-methylpyrrolidone (NMP) without surfactant or polymer additives (both can lower conductivity). NMP has recently been identified to be an excellent solvent for CNTs [8, 11].

Even with optimized CNTnw morphology, the resulting performance was comparable to previous reports, with $R_s$ ∼10–100 kΩ□$^{-1}$ at 98% transparency. Again, the poor conductivity can be attributed to high tube–tube junction resistances, especially between metallic and semiconducting tubes due to Schottky barriers that can be 200 kΩ to greater than 1 MΩ, few methods for mitigating the largest source of resistance, CNT/CNT junctions, have been reported [12–14].

A previous approach demonstrated that individual CNT/CNT junctions can be welded together to form a molecular junction upon irradiation with energetic particles in vacuum at high temperatures (>700°C). Simulations have predicted an increase in electrical conductivity and ductility of the junction upon welding [15, 16]. Experimentally, the increase in electrical conductivity of the resulting nanowelded junctions has yet to be rigorously characterized, although the conductivity was shown to increase in CNT "buckypaper" upon irradiation [16]. In a chemical approach, covalent attachement of gold at CNT/CNT junctions was found to enhance conductivity of a single junction [17]. These studies are interesting but may not be practical for large-area films. Instead we sought to decrease the junction resistance over large areas by nanosoldering the CNT junctions together with a conductive material (Fig. 7.3).

In this work, thin films of fullerenes were deposited onto CNTnws. Fullerenes were chosen due to their electronegativity and strong interaction energies with CNTs. The binding energy of $C_{60}$ at highly coordinated CNT sites (like junctions)

**Fig. 7.3** Optimized CNTnw deposited without surfactants or polymers from NMP (image courtesy of Dr. Melbs LeMieux). The sheet resistance of such films are in $\sim 10$–$100$ k$\Omega\square^{-1}$ range. The image is 25 $\mu$m$^{-2}$

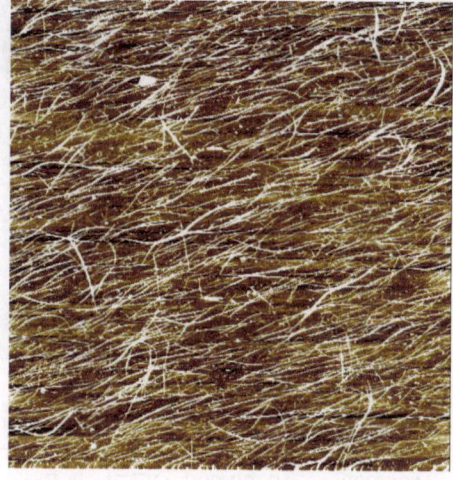

**Fig. 7.4** Schematic of potential energy landscape depicting interactions between a fullerene and a CNT junction. The most energetically favourable location for nucleation is at the junction since the barrier to nucleation is the lowest. The cluster is stabilized by the high coordination and interaction energy

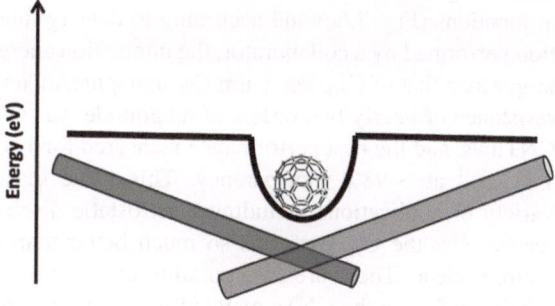

is $\sim 1.0$ and $0.2$–$0.5$ eV along CNT sidewalls, suggesting that nucleation initiates at junctions and then propagates along the CNT (Fig. 7.4) [18]. The strong interaction energies between fullerene and CNTs also ensure that the vast majority of the $C_{60}$ will tend to nucleate and grow on the CNT as opposed to the underlying glass or PET substrate [18, 19]. This is critical, since the interstitial areas between CNTs in the network needs to remain open for high transparency. In short, the idea here is to take advantage of the nucleation theory presented earlier. If given enough thermal energy to diffuse along the surface, the $C_{60}$ (or $C_{70}$) molecules will tend to nucleate initially at the CNT junctions. These sites are the most energetically favourable since the coordination is the greatest and during nucleation, the cluster is stabilized.

Upon deposition of a 1 nm thick layer (all thicknesses quoted are nominal as measured by quartz microbalance), $C_{60}$ preferentially nucleated at junction sites, compared to tube sites. The thin film morphology had small crystallites on the nanotube networks, with the vast majority of these fullerene crystallites occurring at the junctions (Fig. 7.5), implying nucleation is indeed energetically favoured at these sites.

**Fig. 7.5** Scanning electron
micrograph (SEM) showing
the preferential growth of $C_{60}$
(the *white nodes*) at the
junctions sites between CNTs

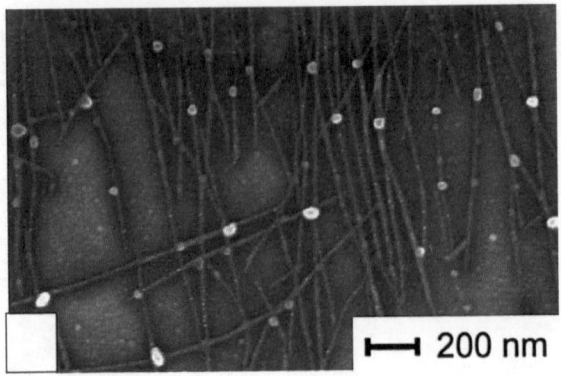

⊢—⊣ 200 nm

After depositing 1 nm of $C_{60}$, the $R_s$ of the network decreased nearly 5–10 times, with a minimal (1% decrease in transparency). The sheet resistance dropped from $\sim 100{,}000$ to $10–20{,}000$ $\Omega\square^{-1}$. indicating that the $C_{60}$ nanosolder is in fact decreasing the junction resistances. $C_{70}$ also showed similar preferential nucleation at junctions (Fig. 7.6c) and according to density functional theory (DFT) calculation performed by a collaborator, the interaction energy of $C_{70}$ on a CNT is 10 meV larger than that of $C_{60}$. For 1 nm $C_{70}$ nanoglue, an unprecedented decrease in sheet resistance of nearly two orders of magnitude was observed relative to unmodified CNTnws, and the best performance measured for 1 nm $C_{70}$ modified CNTnws was 1 k$\Omega\square^{-1}$ at $\sim 98\%$ transparency. This value makes these films suitable for a variety of applications including electrostatic discharge and antistatic films. The reason why the $C_{70}$ performs so much better than the $C_{60}$ as a nanoglue is not entirely clear. There are two possible explanations. The slightly higher binding affinity of $C_{70}$ to the CNTs may aid in charge transport (potentially better overlap between the molecular orbitals of the fullerene and the CNT). Also, the Fermi energy of $C_{70}$ is lower than $C_{60}$, so it is possible the $C_{70}$ is not only acting as a nanoglue, but may also be locally doping the CNTs at the junction (see Fig. 7.6). The fullerene doping is not a conventional interstitial doping, but instead is a charge transfer from the CNT to the fullerene. The charge transfer of an electron from the CNT to $C_{70}$, leaves a mobile hole in the CNT. Figure 7.6 shows the molecular energy levels for $C_{60}$, $C_{70}$, the approximate $C_{60}$ and $C_{70}$ fullerite energy levels, and the range of energy levels for the semiconducting CNTs used. The Fermi energy ($E_f$) values of intrinsic, undoped, semiconductors like fullerenes used in this study is approximately halfway between the highest occupied molecular orbital ($E_{HOMO}$) and lowest unoccupied molecular orbital ($E_{LUMO}$) [$E_f = (E_{HOMO} + E_{LUMO})/2$]. For intrinsic (undoped) semiconductors the Fermi energy is related to the molecular orbital energies, temperatures ($T$), Boltzmann's constant ($k$) and the density of states of the valence ($N_v$) and conduction bands ($N_c$):

$$E_F = \frac{E_{HOMO} + E_{LUMO}}{2} + \frac{kT}{2}\ln\left(\frac{N_v}{N_c}\right) \qquad (7.1)$$

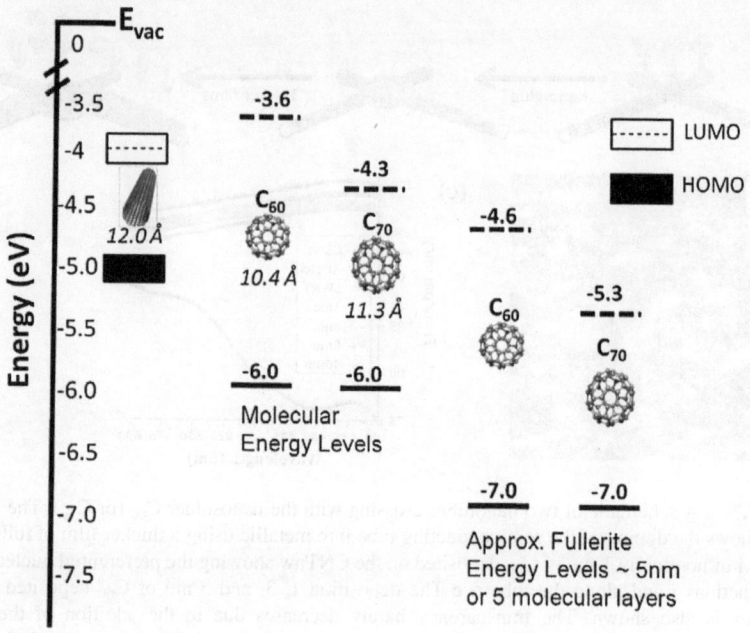

**Fig. 7.6** The molecular orbital energies of the CNTs and fullerenes. For the CNTs there are a wide range of HOMO and LUMO energies which are represented by the band of energies The energy levels can drop nearly 1 eV when solid fullerite is formed. The approximate conduction and valence band energies for $C_{60}$ and $C_{70}$ fullerite are also given

The HOMO and LUMO energy values are much larger than the thermal energy $kT$ (0.026 eV) and the natural log function also decreases the value of the second term on the right hand side of the equation, so that the Fermi energy can safely be approximated as halfway between the HOMO and LUMO energies. For fullerenes, the Fermi energy ($E_f$) can decreases nearly 1 eV as fullerite ($\sim 5$ monolayers) is formed; the $E_f$s of solid $C_{60}$ or $C_{70}$ can be $\sim 1$ eV lower than molecular $C_{60}$ or $C_{70}$, so thicker films of both $C_{60}$ and $C_{70}$ can potentially dope CNTs (Fig. 7.6) [19]. In order to test this hypothesis, 5 nm films of $C_{60}$ and $C_{70}$ were deposited onto the CNTnws.

After 5 nm of $C_{60}$ was deposited, a bamboo-like structure formed on the entire CNTs. The preference for the $C_{60}$ on the CNT compared to the substrate again is due to favourable interaction energies. Concomitant with the AFM and SEM micrographs, the preference for growth on the CNTnw is evident from the UV–Vis measurements, which show that the transparency of the hybrid film barely decreases (<3%) even after 5 nm of fullerene was deposited (Fig. 7.7). The amount of material on the network increased after deposition, while the transparency has essentially remained the same.

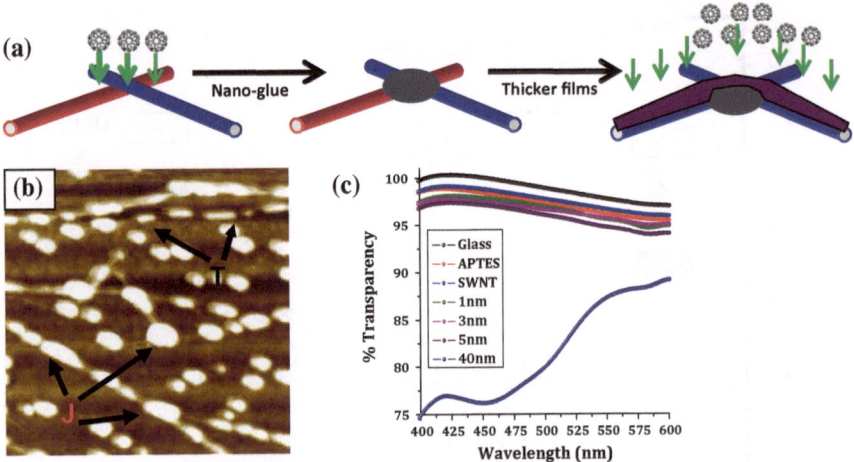

**Fig. 7.7** **a** A schematic of two nanotubes crossing with the nanosolder $C_{60}$ (or $C_{70}$). The second step shows the doping of the semiconducting tube into metallic using a thicker film of fullerenes. **b** AFM of nominally 1 nm of $C_{60}$ deposited on the CNTnw showing the preferential nucleation at the junctions $j$ and along the tube $t$. **c** The deposition 1, 3, and 5 nm of $C_{60}$ deposited on the CNTnw is also shown. The transparency barely decreases due to the addition of the small molecule organic semiconductors. Finally, only after 40 nm of $C_{60}$ is deposited does the transparency decrease substantially

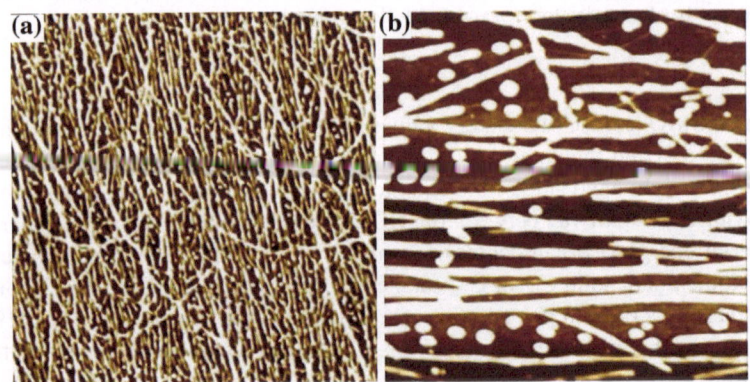

**Fig. 7.8** **a** AFM of a dense CNTnw after 5 nm of $C_{70}$ has been deposited (area is 100 $\mu m^2$). **b** A zoomed in, less dense CNTnw after 5 nm of $C_{70}$ has been deposited (area is 4 $\mu m^2$). The conformal growth and tight binding is clearly visible. The scale for both images is 20 nm

$C_{70}$ exhibited more conformal growth on the nanotubes than $C_{60}$. Rather than the bamboo-like structure observed with $C_{60}$, a continuous thinner "nanowire" morphology was observed after a 5 nm deposition of $C_{70}$ (Fig. 7.8).

The deposition of thicker films of the fullerenes resulted in even lower sheet resistances. CNTnws with 5 nm of $C_{70}$ were found to be much more conductive

**Fig. 7.9** The effect of $C_{60}$ and $C_{70}$ deposition on CNTnws. After deposition of 5 nm of $C_{70}$, the sheet resistance of the CNTnw decreased nearly two orders of magnitude

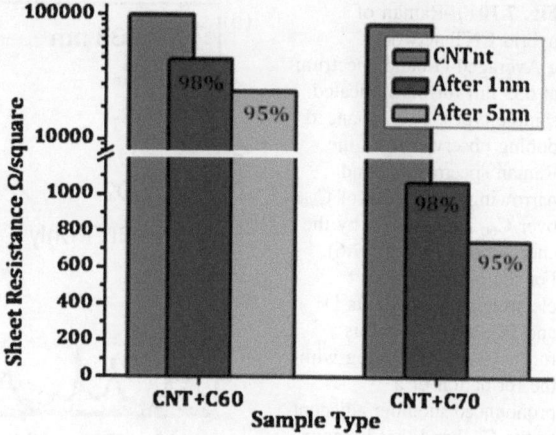

than similar networks on which $C_{60}$ had been deposited (Fig. 7.9). CNTnws with 5 nm of $C_{70}$ showed sheet resistances as low as 600 $\Omega\square^{-1}$ at 95% transparency.

From the thicker films of $C_{70}$ deposited, it became evident that the increase in conductivity was not only due to improvement of junction resistances, but also due to doping of the semiconducting tubes. When the fullerenes and CNTs come into contact, and equilibrate, the Fermi energy of the two materials must also become equal. This occurs when the higher energy electrons in the CNT highest molecular orbital (HOMO) populate the lowest molecular orbital (LUMO) energies of the fullerenes. This doping becomes prominent when fullereite is formed since the LUMO for $C_{70}$ decreases substantially [19]. From this result, it is hypothesized that the decrease in CNTnw sheet resistance due to deposition of $C_{60}$ is primarily due to the morphological enhancement that provides increased area for charge transport, and intimate contact at CNT junctions. However, for $C_{70}$, a combined gluing and doping effect occurs. The more conformal growth with $C_{70}$ and the lower $E_f$ compared to $C_{60}$ resulted in much more effective doping, since the number of holes introduced into the CNTnw is directly related to the area of contact between the fullerene and the CNT. The $E_f$ of $C_{60}$ is comparable (Fig. 7.6) to the $E_{HOMO}$ of the semiconducting CNTs (roughly $-4.0$ eV for the arc-discharge tubes used here), therefore electron transfer to $C_{60}$ is possible, but not nearly as favourable as to $C_{70}$ (confirmed by $\mu$-Raman spectroscopy).

Finally, in addition to using energy level diagrams, charge transfer doping is verified with micro($\mu$)-Raman spectroscopy. $\mu$-Raman spectra were obtained for hybrid samples (with nominally 5 nm of $C_{60}$, or $C_{70}$ deposited on CNTnw) using 633 and 532 nm laser excitations. Investigating the tangential mode region (G-band) at 633 nm excitation for the $C_{60}$ and $C_{70}$ nanoglue dopant, prominent differences are observed (Fig. 7.10) for the $C_{70}$/CNT hybrid compared to the $C_{60}$/CNT hybrid [20, 21]. Other dopants with even lower molecular energy levels were also deposited to confirm charge transfer. The $C_{60}$/CNT hybrid spectrum

**Fig. 7.10** $\mu$-Raman of hybrid CNT networks. **a** Averaged G-band spectrum at 633 nm for the indicated sample types. The enhanced doping observed from the Raman spectra $G^-$ band narrowing in the case of $C_{70}$ over $C_{60}$ is explained by the energy diagram (Fig. 7.6). For the other more electronegative dopants D1 and D2, the $G^-$ band is further narrowed, along with the formation of a pronounced shoulder adjacent to the $G^-$ band that is more evident in the $C_{60}$ $F_{48}$ case. As noted in the text, the formation of peapods is unlikely since only $C_{60}$ has a small enough diameter for this to occur. For $C_{60}$ there are no evident peapod peaks observed in the $\mu$-Raman spectra. **b** Averaged RBM spectrum at 633 nm for the CNT networks hybridized with the indicated fullerene, compared with a bare CNT network. Here, the RBM is quenched with the level of doping as indicated in (**a**). All spectra were normalized, and taken from films with 5 nm of $C_{60}$ and $C_{70}$. Image courtesy of Dr. Melbs LeMieux

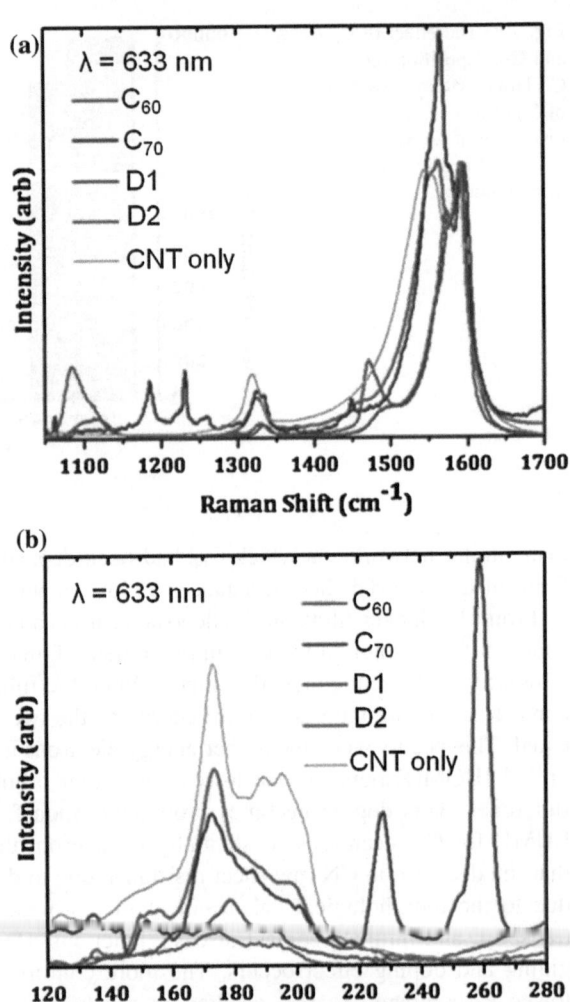

resembles a pristine CNTnw in terms of lack of $G^-$ band position and lack of non-CNT peaks over the entire spectrum. The pronounced narrowing of $G^-$ peak (Fig. 7.10) in the $C_{70}$/CNT hybrid relative to a bare CNTnw (and the $C_{60}$/CNTnw) has previously been attributed to charge transfer. Correspondingly, the characteristic radial breathing mode (RBM) undergoes a loss in intensity for the $C_{70}$/CNTnw. This reflects a much larger degree of charge transfer compared to the $C_{60}$/CNTnw, and validates the $C_{70}$ usage as a more efficient nanoglue dopant [20, 21].

These results indicate that charge transfer from the CNT to the dopant increases as from $C_{60}$ to $C_{70}$ (i.e. increases as small molecule $E_{LUMO}$ decreases) and are commensurate with the conductivity trend observed upon doping the CNTnws.

$\mu$-Raman spectroscopy can also verify the lack of peapod formation [22, 23]. For the fullerenes used in this work, the smallest ($C_{60}$) is the most likely to form the peapod structure in the $\sim 1.5$ nm diameter arc-discharge SWNTs. However, due to the relatively fast deposition rates (0.3 Å s$^{-1}$), low defect density of the CNTs (small peak at 1,325 cm$^{-1}$ defect band, Fig. 7.10a), and low substrate temperature used during deposition, peapod formation should be suppressed. This is confirmed by analysis of the RBM. RBM peak positions remain unchanged after the $C_{60}$ deposition, while for peapods a redshift of the 173 (633) and 188 cm$^{-1}$ (532 nm) was reported [22, 23]. $C_{70}$ peapods should show a splitting in the 260 cm$^{-1}$ mode at 532 nm excitation which again was not observed [22, 23].

## 7.5 Conclusions

Using $C_{60}$ and $C_{70}$ that preferentially grow on CNTs and CNT junctions, the most energetically favorable site for nucleation, as a nanosolder, and then further doping the film with thicker fullerene films, a high performance carbon based TE has been fabricated. Deposition of fullerenes morphologically and electronically glued the partially aligned CNTnws together, thereby decreasing junction resistance, while thicker fullerene films further decreased the overall CNTnw resistance. In doing so, major hurdles with regard to CNT-based transparent electrodes were overcome by focusing on the most critical issues: decreasing the junction resistance and chirality mixture in these films. Furthermore, the ultrathin hybrid CNTnw is stable for months in ambient and at 250 °C for over 40 h, fabricated using simple processing techniques at low temperatures, and compatible with flexible substrates making it a more attractive ITO replacement than other technologies. In addition to producing a new carbon-based high performance transparent electrode material, this new large-area nanoglue concept, which employs selective nucleation at junctions, may be an effective approach for forming and connecting nanoassemblies in parallel that may yield exciting materials for applications in opto- and nanoelectronics.

## 7.6 Experimental

### 7.6.1 Materials and Methods

#### Preparation of Surfaces

Glass substrates from corning were cleaned for 45 min in hot Piranha solution (3:1 $H_2SO_4$:$H_2O_2$), rinsed copiously with water, dried under $N_2$, and taken inside a dry $N_2$ glovebox for silane modification. SAMs of aminopropyltriethoxy silane (APTES, purchased from Gelest Inc, silanes were distilled prior to use) were formed from 0.4% solution (in anhydrous toluene) for 1 h at room temperature. The surfaces were rinsed twice in toluene, sonicated in toluene, and rinsed again in toluene, then dried under $N_2$. With these conditions, the substrates typically have contact angle of 60° and ellipsometry measured thickness of 0.6 nm. The $C_{60}$ (99%) and $C_{70}$ (99%) fullerenes (SES research) were used as received.

**Nanotube Solution Preparation**

Solutions of the purified arc-discharge nanotubes (see previous reference for purification details) were dispersed by sonication for 45 min (solution kept in ice bath) in NMP (1-methyl-2-pyrrolidone, Omnisolve, Spectrophotometry grade) at concentrations ranging from 10 to 500 $\mu g\ ml^{-1}$.

**Sample Fabrication**

Typically, 400–600 $\mu l$ of the CNT solution was spincoated onto the APTES treated display glass at RPMs ranging from 1,000 to 4,000 RPM to achieve partial SWNT alignment. Afterwards, the glass was heated under vacuum to remove residual solvent. 40 nm gold electrodes with 2 mm spacing were then deposited (thermal evaporation) for characterization of the sheet resistance. Finally, fullerene deposition took place via thermal evaporation with thicknesses ranging from 1–40 nm, at a pressure of $10^{-6}$ Torr. Fullerenes were also deposited form solution (1–5 mg $ml^{-1}$) in toluene or NMP.

**Sample Characterization**

AFM topography images were acquired in the tapping mode regime using a Multimode AFM (Veeco). All electronic tests were conducted using a Keithley 4200 SC semiconductor parameter analyzer. Micro-Raman (LabRam Aramis, Horiba Jobin-Yvon) measurements were carried out at 633 (1.96) and 532 nm (2.41 eV) excitations at 100x magnification and 1 um spot size, and a 1,200 grating. Excitation power was 2 mW for the 633 nm line and 1 mW for the 532 nm line. All data was acquired from automated multi-point (usually 12 points) mapping over random areas (three different areas, excluding the extreme 2 mm diameter center of the wafer) of the samples, with two spectra accumulated and averaged at each single point. UV–Vis (Cary 6000i) measurements were done at 300 nm $min^{-1}$ with 1 nm wavelength intervals. SEM images were collected using a Raith 150 (Gm bH) without gold coating at low operating voltages (3 kV) using a secondary electron detector.

# References

1. Gruner G (2006) Carbon nanotube films for transparent and plastic electronics. J Mater Chem 16:3533–3539
2. Tung VC, Chen LM, Allen MJ, Wassei JK, Nelson K, Kaner RB, Yang Y (2009) Low-temperature solution processing of graphene–carbon nanotube hybrid materials for high-performance transparent conductors. Nano Lett 9(5):1949–1955
3. Lee JY, Connor ST, Cui Y, Peumans P (2008) Solution-processed metal nanowire mesh transparent electrodes. Nano Lett 8(2):689–692
4. Kang MG, Guo LJ (2007) Nanoimprinted semitransparent metal electrodes and their application in organic light-emitting diodes. Adv Mater 19:1391
5. Zhang M, Fang SL, Zakhidov AA, Lee SB, Aliev AE, Williams CD, Atkinson KR, Baughman RH (2005) Strong, transparent, multifunctional, carbon nanotube sheets. Science 309(5738):1215–1219
6. Wu ZC, Chen ZH, Du X, Logan JM, Sippel J, Nikolou M, Kamaras K, Reynolds JR, Hebard Tanner DB, AF Rinzler AG (2004) Transparent, conductive carbon nanotube films. Science 305(5688):1273–1276
7. Gu H, Swager TM (2008) Fabrication of free-standing, conductive, and transparent carbon nanotube films. Adv Mater 20(23):4433–4437
8. LeMieux MC, Roberts M, Barman S, Jin YW, Kim JM, Bao Z (2008) Self-sorted, aligned nanotube networks for thin-film transistors. Science 321(5885):101–4
9. Hellstrom SL, Lee HW, Bao ZN (2009) Polymer-assisted direct deposition of uniform carbon nanotube bundle networks for high performance transparent electrodes. Acs Nano 3(6):1423–1430
10. Topinka MA, Rowell MW, Goldhaber-Gordon D, McGehee MD, Hecht DS, Gruner G (2009) Charge transport in interpenetrating networks of semiconducting and metallic carbon nanotubes. Nano Lett 9(5):1866–1871
11. Bergin SD, Nicolosi V, Streich PV, Giordani S, Sun ZY, Windle AH, Ryan P, Niraj NPP, Wang ZTT, Carpenter L, Blau WJ, Boland JJ, Hamilton JP, Coleman JN (2008) Towards solutions of single-walled carbon nanotubes in common solvents. Adv Mater 20(10):1876
12. Bachtold A, Fuhrer MS, Plyasunov S, Forero M, Anderson EH, Zettl A, McEuen PL (2000) Scanned probe microscopy of electronic transport in carbon nanotubes. Phys Rev Lett 84(26):6082–6085
13. Nirmalraj PN, Lyons PE, De S, Coleman JN, Boland JJ (2009) Electrical connectivity in single-walled carbon nanotube networks. Nano Lett
14. Terrones M, Banhart F, Grobert N, Charlier JC, Terrones H, Ajayan PM (2002) Molecular junctions by joining single-walled carbon nanotubes. Phys Rev Lett 89(7):075505
15. Jang I, Sinnott SB, Danailov D, Keblinski P (2003) Molecular dynamics simulation study of carbon nanotube welding under electron beam irradiation. Nano Lett 4(1):109–114
16. Ishaq A, Yan L, Zhu D (2009) The electrical conductivity of carbon nanotube sheets by ion beam irradiation. Nucl Instr Methods Phys Res Sect B Beam Interact Mat Atoms 267(10):1779–1782
17. Velamakanni A, Magnuson CW, Ganesh KJ, Zhu Y, An J, Ferreira PJ, Ruoff RS. Site-specific deposition of au nanoparticles in CNT films by chemical bonding. ACS Nano
18. Ulbricht H, Moos G, Hertel T (2003) Interaction of C-60 with carbon nanotubes and graphite. Phys Rev Lett 90(9)
19. Girifalco LA, Hodak M, Lee RS (2000) Carbon nanotubes, buckyballs, ropes, and a universal graphitic potential. Phys Rev B 62(19):13104
20. McGuire K, Gothard N, Gai PL, Dresselhaus MS, Sumanasekera G, Rao AM (2005) Synthesis and Raman characterization of boron-doped single-walled carbon nanotubes. Carbon 43(2):219–227
21. Nasibulin AG, Pikhitsa PV, Jiang H, Brown DP, Krasheninnikov AV, Anisimov AS, Queipo P, Moisala A, Gonzalez D, Lientschnig G, Hassanien A, Shandakov SD, Lolli G, Resasco

DE, Choi M, Tomanek D, Kauppinen EI (2007) A novel hybrid carbon material. Nat Nano 2(3):156–161
22. Pichler T, Kuzmany H, Kataura H, Achiba Y (2001) Metallic polymers of C60 inside single-walled carbon nanotubes. Phys Rev Lett 87(26):267401
23. Kavan L, Dunsch L, Kataura H, Oshiyama A, Otani M, Okada S (2003) Electrochemical tuning of electronic structure of C60 and C70 fullerene peapods: in situ visible near-infrared and Raman study. J Phys Chem B 107(31):7666–7675

# Chapter 8
# Outlook/Conclusions

## 8.1 Outlook

Overall there still remains several technical challenges which face the widespread utilization of organic electronics. The stability and performance or organic semiconductors still needs improvement. There are also still issues associated with robust and reliable patterning. Nevertheless, the potential for low-cost, flexible, and printable electronics is so attractive that organic electronics will remain an important scientific and engineering focus for academia and industry. Moreover, the discovery of new high performance carbon based materials like CNTs and graphene will further research efforts in plastic electronics since these materials have conductivities and mechanical properties far exceeding conventional inorganics.

Based on the work presented in this thesis, several future directions may be interesting. First, fabrication of organic circuits using the crystalline octadecylsilane monolayer requires further developments in patterning and processing. Experiments exploring the kinetics of spin-cast alkylsilane monolayer fabrication using grazing incidence X-ray diffraction are on-going. It will also be interesting to study the odd–even effect on organic semiconductor nucleation and growth on alkylsilane monolayers. The development of low voltage monolayer transistors and sensors is another promising area. Finally, studying the charge transport of CNTnws with nanoglue is an on-going project. There is interest in investigating their thin film structure using X-ray diffraction.

## 8.2 Conclusions

In summary, this thesis has resulted in the following findings which should be valuable to the development of organic electronics:

A. Virkar, *Investigating the Nucleation, Growth, and Energy Levels of Organic Semiconductors for High Performance Plastic Electronics*, Springer Theses, DOI: 10.1007/978-1-4419-9704-3_8, © Springer Science+Business Media, LLC 2012

1. The importance of controlling and engineering organic semiconductor growth for high performance electronics has been shown. Promoting the two-dimensional growth of organic semiconductors at the dielectric interface leads to very high charge carrier mobilities.
2. A systematic study of the nucleation and growth of organic semiconductors on the most common surface used for organic transistors has been presented. The energetics of pentacene on octadecylsilane modified surfaces were analyzed in depth. Numerical heuristics and interaction energies necessary to drive desirable growth were calculated. Also, the energetics of nucleation were studied and it was determined that the density of octadecylsilane monolayer affects the barrier to nucleation and not diffusivity as was previously suggested.
3. The general importance of a crystalline monolayer of octadecylsilane for organic transistors was demonstrated. The development of a spin-casting crystalline octadecylsilane provides a simple and scalable method for producing surfaces which may have important technological implications. The crystalline octadecylsilane monolayer allowed for the ability to fabricate low voltage monolayer transistors which may very interesting for both organic circuits and in the emerging field of organic transistor based sensors.
4. The selective nucleation and growth of organic small molecules on carbon nanotube networks lead to record performance transparent electrodes. The nucleation of fullerenes on the nanotube junctions in a network greatly reduced the junction resistances, while thicker fullerene films provide stable and efficient doping. Prototype solar cells and touch screens have been fabricated using the new hybrid carbon based electrode.